The Craft of Fractional Modelling in Science and Engineering

Special Issue Editor

Jordan Hristov

MDPI • Basel • Beijing • Wuhan • Barcelona • Belgrade

Special Issue Editor
Jordan Hristov
University of Chemical Technology and Metallurgy
Bulgaria

Editorial Office
MDPI
St. Alban-Anlage 66
Basel, Switzerland

This edition is a reprint of the Special Issue published online in the open access journal *Fractal Fract* (ISSN 2504-3110) from 2017–2018 (available at: http://www.mdpi.com/journal/fractalfract/special_issues/Fractional_Modelling).

For citation purposes, cite each article independently as indicated on the article page online and as indicated below:

Lastname, F.M.; Lastname, F.M. Article title. *Journal Name* **Year**, *Article number*, page range.

First Edition 2018

ISBN 978-3-03842-983-8 (Pbk)
ISBN 978-3-03842-984-5 (PDF)

Articles in this volume are Open Access and distributed under the Creative Commons Attribution (CC BY) license, which allows users to download, copy and build upon published articles even for commercial purposes, as long as the author and publisher are properly credited, which ensures maximum dissemination and a wider impact of our publications. The book taken as a whole is © 2018 MDPI, Basel, Switzerland, distributed under the terms and conditions of the Creative Commons license CC BY-NC-ND (http://creativecommons.org/licenses/by-nc-nd/4.0/).

Table of Contents

About the Special Issue Editor

Jordan Hristov is a professor of Chemical Engineering with the University of Chemical Technology and Metallurgy, Sofia, Bulgaria. He graduated from the Technical University, Sofia, as an electrical engineer and was awarded his Ph.D. degree in chemical engineering in 1994 with a thesis in the field of magnetic field assisted fluidization. His Doctor of Sciences thesis on nonlinear and anomalous diffusion models was successfully completed in 2018. Prof. Hristov has more than 39 years' experience in the field of chemical engineering with principle research interests in mechanics of particulate materials, fluidization, magnetic field effects of process intensification, mathematical modelling in complex systems, non-linear diffusion and fractional calculus application in modelling with more than 170 articles published in international journals. He is an editorial board member of Thermal Science, Particuology, Progress in Fractional Differentiation and Applications and Fractional and Fractal.

Preface to "The Craft of Fractional Modelling in Science and Engineering"

Fractional calculus has performed an important role in the fields of mathematics, physics, electronics, mechanics, and engineering in recent years. The modeling methods involving fractional operators have been continuously generalized and enhanced, especially during the last few decades. Many operations in physics and engineering can be defined accurately by using systems of differential equations containing different types of fractional derivatives.

This book is a result of the contributions of scientists involved in the special collection of articles organized by the journal *Fractal and Fractional* (MDPI), most of which have been published at the end of 2017 and the beginning 2018. In accordance with the initial idea of a Special Issue, the best published have now been consolidated into this book.

The articles included span a broad area of applications of fractional calculus and demonstrate the feasibility of the non-integer differentiation and integration approach in modeling directly related to pertinent problems in science and engineering. It is worth mentioning some principle results from the collected articles, now presented as book chapters, which make this book a contemporary and interesting read for a wide audience:

The fractional velocity concept developed by Prodanov [1] is demonstrated as tool to characterize Hölder and in particular, singular functions. Fractional velocities are defined as limits of the difference quotients of a fractional power and they generalize the local notion of a derivative. On the other hand, their properties contrast some of the usual properties of derivatives. One of the most peculiar properties of these operators is that the set of their nontrivial values is disconnected. This can be used, for example, to model instantaneous interactions, such as Langevin dynamics. In this context, the local fractional derivatives and the equivalent fractional velocities have several distinct properties compared to integer-order derivatives.

The classical pantograph equation and its generalizations, including fractional order and higher order cases, is developed by Bhalekar and Patade [2]. The special functions are obtained from the series solution of these equations. Different properties of these special functions are established, andtheir relations with other functions are developed.

The new direction in fractional calculus involving nonsingular memory kernels, developed in the last three years following the seminar articles of Caputo and Fabrizio in 2015 [3], is hot research topic. Two studies in the collection clearly demonstrate two principle directions: operators with nonsingular exponential kernels, i.e., the so-called Caputo-Fabrizio derivatives [4,5] (Hristov, 2016, Chapter 10), and operators with nonsingular memory kernels based on the Mittag-Leffler function [6,7] (Atangana, Baleanu, 2016; Baleanu, Fennandez, 2018).

Yavuz and Ozdemir [8] demonstrate a novel approximate-analytical solution method, called the Laplace homotopy analysis method (LHAM), using the Caputo-Fabrizio (CF) fractional derivative operator based on the exponential kernel. The recommended method is obtained by combining Laplace transform (LT) and the homotopy analysis method (HAM). This study considers the application of LHAM to obtain solutions of the fractional Black-Scholes equations (FBSEs) with the Caputo-Fabrizio (CF) fractional derivative and appropriate initial conditions. The authors demonstrate the efficiencies and accuracies of the suggested method by applying it to the FBS option pricing models with their initial conditions satisfied by the classical European vanilla option. Using real market values from finance literature, it is demonstrated how the option is priced for fractional cases of European call option pricing models. Moreover, the proposed fractional model allows modeling of the price of different financial derivatives such as swaps, warrant, etc., in complete agreement with the corresponding exact solutions.

In light of new fractional operators, Gomez-Aguilar and Atangana [9] present alternative representations of the Freedman model considering Liouville-Caputo and Atangana-Baleanu-Caputo fractional derivatives. The solutions of these alternative models are obtained using an iterative scheme based on the Laplace transform and the Sumudu transform. Moreover, special solutions via the Adams-Moulton rule are obtained for both fractional derivatives.

In light of certain applied problems, the thermal control of complex thermal interfaces and heat conduction are principle issues to which classical fractional calculus is widely applicable. The control of thermal interfaces has gained importance in recent years because of the high cost of heating and cooling materials in many applications. The main focus in the work of Moreau et al. [10] is to compare the second and third generations of the CRONE controller (French acronym of Commande Robusted'Ordre Non Entier) in the control of a non-integer plant by means a fractional order controller. The idea is that, as a consequence of the fractional approach, all of the systems of integer order are replaced by the implementation of a CRONE controller. The results reveal that the second generation CRONE controller is robust when the variations in the plant are modeled with gain changes, whereas the phase remains the same for all of the plants (even if not constant). However, the third generation CRONE controller demonstrates a good, feasible robustness when the parameters of the plant are changed as well as when both gain and phase variations are encountered.

Thermistors are part of a larger group of passive components. They are temperature-dependent resistors and come in two varieties, negative temperature coefficients (NTCs) and positive temperature coefficients (PTCs), although NTCs are most commonly u sed. NTC thermistors are nonlinear, and their resistance decreases as the temperature increases. The self-heating may affect the resistance of an NTC thermistor, and the work of Vivek et al. [11] focuses on the existence and uniqueness and Ulam-Hyers stability types of solutions for Hilfer-type thermistor problems.

A heat conduction inverse problem of the fractional (Caputo fractional derivative) heat conduction problem is developed by Brociek et al. [12] for porous aluminum. In this case, the Caputo fractional derivative is employed. The direct problem is solved using a finite difference method and approximations of Caputo derivatives, while the inverse problem, the heat transfer coefficient, thermal conductivity coefficient, initial condition, and order of derivative are sought and the minimization of the functional describing the error of approximate solution is carried out by the Real Ant Colony Optimization algorithm.

Process monitoring represents an important and fundamental tool aimed at process safety and economics while meeting environmental regulations. Lenzi et al. [13] present an interesting approach to the quality control of different olive and soybean oil mixtures characterized by image analysis with the aid of an RGB color system by the algebraic fractional model. The model based on the fractional calculus-based approach could better describe the experimental dataset, presenting better results of parameter estimation quantities, such as objective function values and parameter variance. This model could successfully describe an independent validation sample, while the integer order model failed to predict the value of the validation sample.

The classical Stokes' first problem for a class of viscoelastic fluids with the generalized fractional Maxwell constitutive model was developed by Bazhlekova and Bazhlekov [14]. The constitutive equation is obtained from the classical Maxwell stress-strain relation by substituting the first-order derivatives of stress and strain by derivatives of non-integer orders in the interval $(0, 1)$. Explicit integral representation of the solution is derived and some of its characteristics are discussed: non-negativity and monotonicity, asymptotic behavior, analyticity, finite/infinite propagation speed, and absence of wave front.

Summing up, this special collection presents a detailed picture of the current activity in the field of fractional calculus with various ideas, effective solutions, new derivatives, and solutions to applied problems. As editor, I believe that this will be continued as series of Special Issues and books released to further explore this subject.

I would like to expresses my gratitude to Mrs. Colleen Long from the office of *Fractal and Fractional* for the correct and effective work in handling the submitted manuscripts. The work and comments of the reviewers allowing this collection to be published are gratefully acknowledged as well.

Last but not least, the efforts of all authors contributing to this special collection arehighly appreciated.

Conflicts of Interest: The author declares no conflict of interest.

Jordan Hristov
Special Issue Editor

References

1. Prodanov, D. Fractional Velocity as a Tool for the Study of Non-Linear Problems. *Fractal Fract.* **2018**, *2*, 4. [CrossRef]
2. Bhalekar, S.; Patade, J. Series Solution of the Pantograph Equation and Its Properties. *Fractal Fract.* **2017**, *1*, 16. [CrossRef]
3. Caputo, M.; Fabrizio, M. A new definition of fractional derivative without singular kernel. *Prog. Fract. Differ. Appl.* **2015**, *1*, 1–13. [CrossRef]
4. Hristov, J. Transient heat diffusion with a non-singular fading memory: From the Cattaneo constitutive equation with Jeffrey's kernel to the Caputo-Fabrizio time-fractional derivative. *Therm. Sci.* **2016**, *20*, 757–762. [CrossRef]
5. Hristov, J. Fractional derivative with non-singular kernels from the Caputo-Fabrizio definition and beyond: Appraising analysis with emphasis on diffusion models. In *Frontiers in Fractional Calculus*; Bhalekar, S., Ed.; Bentham Science Publishers: Sharjah, UAE, 2017; pp. 269–342.
6. Atangana, A.; Baleanu, D. New fractional derivatives with non-local and non-singular kernel: Theory and application to Heat transfer model. *Therm. Sci.* **2016**, *20*, 763–769. [CrossRef]
7. Baleanu, D.; Fernandez, A. On some new properties of fractional derivatives with Mittag-Leffler kernel. *Commun. Nolinear Sci. Numer. Simul.* **2018**, *59*, 444–462. [CrossRef]
8. Yavuz, M.; Özdemir, N. European Vanilla Option Pricing Model of Fractional Order without Singular Kernel. *Fractal Fract.* **2018**, *2*, 3. [CrossRef]
9. Gómez-Aguilar, J.F.; Atangana, A. Fractional Derivatives with the Power-Law and the Mittag–Leffler Kernel Applied to the Nonlinear Baggs–Freedman Model. *Fractal Fract.* **2018**, *2*, 10. [CrossRef]
10. Moreau, X.; Daou, R.A.Z.; Christophy, F. Comparison between the Second and Third Generations of the CRONE Controller: Application to a Thermal Diffusive Interface Medium. *Fractal Fract.* **2018**, *2*, 5. [CrossRef]
11. Vivek, D.; Kanagarajan, K.; Sivasundaram, S. Dynamics and Stability Results for Hilfer Fractional Type Thermistor Problem. *Fractal Fract.* **2017**, *1*, 5. [CrossRef]
12. Brociek, R.; Słota, D.; Król, M.; Matula, G.; Waldemar Kwasny, W. Modeling of Heat Distribution in Porous Aluminum Using Fractional Differential Equation. *Fractal Fract.* **2017**, *1*, 17. [CrossRef]
13. Lenzi, E.K.; Ryba, A.; Lenzi, M.K. Monitoring Liquid-Liquid Mixtures Using Fractional Calculus and Image Analysis. *Fractal Fract.* **2018**, *2*, 11. [CrossRef]
14. Bazhlekova, E.; Bazhlekov, I. Stokes' First Problem for Viscoelastic Fluids with a Fractional Maxwell Model. *Fractal Fract.* **2017**, *1*, 7. [CrossRef]

fractal and fractional

MDPI

Article

Fractional Velocity as a Tool for the Study of Non-Linear Problems

Dimiter Prodanov

Environment, Health and Safety, IMEC vzw, Kapeldreef 75, 3001 Leuven, Belgium; Dimiter.Prodanov@imec.be or dimiterpp@gmail.com

Received: 27 December 2017; Accepted: 15 January 2018; Published: 17 January 2018

Abstract: Singular functions and, in general, Hölder functions represent conceptual models of nonlinear physical phenomena. The purpose of this survey is to demonstrate the applicability of fractional velocities as tools to characterize Hölder and singular functions, in particular. Fractional velocities are defined as limits of the difference quotients of a fractional power and they generalize the local notion of a derivative. On the other hand, their properties contrast some of the usual properties of derivatives. One of the most peculiar properties of these operators is that the set of their non trivial values is disconnected. This can be used for example to model instantaneous interactions, for example Langevin dynamics. Examples are given by the De Rham and Neidinger's singular functions, represented by limits of iterative function systems. Finally, the conditions for equivalence with the Kolwankar-Gangal local fractional derivative are investigated.

Keywords: fractional calculus; non-differentiable functions; Hölder classes; pseudo-differential operators

MSC: Primary 26A27; Secondary 26A15, 26A33, 26A16, 47A52, 4102

---

1. Introduction

Non-linear and fractal physical phenomena are abundant in nature [1,2]. Examples of non-linear phenomena can be given by the continuous time random walks resulting in fractional diffusion equations [3], fractional conservation of mass [4] or non-linear viscoelasticity [5,6]. Such models exhibit global dependence through the action of the nonlinear convolution operator (i.e., differ-integral). Since this setting opposes the principle of locality there can be problems with the interpretation of the obtained results. In most circumstances such models can be treated as asymptotic as it has been demonstrated for the time-fractional continuous time random walk [7]. The asymptotic character of these models leads to the realization that they describe *mesoscopic* behavior of the concerned systems. The action of fractional differ-integrals on analytic functions results in Hölder functions representable by fractional power series (see for example [8]).

Fractals are becoming essential components in the modeling and simulation of natural phenomena encompassing many temporal or spatial scales [9]. The irregularity and self-similarity under scale changes are the main attributes of the morphologic complexity of cells and tissues [10]. Fractal shapes are frequently built by iteration of function systems via recursion [11,12]. In a large number of cases, these systems leads to nowhere differentiable fractals of infinite length, which may be unrealistic. On the other hand fractal shapes observable in natural systems typically span only several recursion levels. This fact draws attention to one particular class of functions, called *singular*, which are differentiable but for which at most points the derivative vanishes. There are fewer tools for the study of singular functions since one of the difficulties is that for them the Fundamental Theorem of calculus fails and hence they cannot be represented by a non-trivial differential equation.

Singular signals can be considered as toy-models for strongly-non linear phenomena, such as turbulence or asset price dynamics. In the beginning of 1970s, Mandelbrot proposed a model of random energy dissipation in intermittent turbulence [13], which served as one of the early examples of multifractal formalism. This model, known as canonical Mandelbrot cascades, is related to the Richardson's model of turbulence as noted in [14,15]. Mandelbrot's cascade model and its variations try to mimic the way in which energy is dissipated, i.e., the splitting of eddies and the transfer of energy from large to small scales. There is an interesting link between multifractals and Brownian motion [16]. One of the examples in the present paper can be related to the Mandelbrot model and the associated binomial measure.

Mathematical descriptions of strongly non-linear phenomena necessitate relaxation of the assumption of differentiability [17]. While this can be achieved also by fractional differ-integrals, or by multi-scale approaches [18], the present work focuses on local descriptions in terms of limits of difference quotients [19] and non-linear scale-space transformations [20]. The reason for this choice is that locality provides a direct way of physical interpretation of the obtained results. In the old literature, difference quotients of functions of fractional order have been considered at first by du Bois-Reymond [21] and Faber [22] in their studies of the point-wise differentiability of functions. While these initial developments followed from purely mathematical interest, later works were inspired from physical research questions. Cherbit [19] and later on Ben Adda and Cresson [23] introduced the notion of fractional velocity as the limit of the fractional difference quotient. Their main application was the study of fractal phenomena and physical processes for which the instantaneous velocity was not well defined [19].

This work will further demonstrate applications to singular functions. Examples are given by the De Rham and Neidinger's functions, represented by iterative function systems (IFS). The relationship with the Mandelbrot cascade and the associated Bernoulli-Mandelbrot binomial measure is also put into evidence. In addition, the form of the Langevin equation is examined for the requirements of path continuity. Finally, the relationship between fractional velocities and the localized versions of fractional derivatives in the sense of Kolwankar-Gangal will be demonstrated.

2. Fractional Variations and Fractional Velocities

The general definitions and notations are given in Appendix A. This section introduces the concept of fractional variation and fractional velocities.

Definition 1. *Define forward (backward) Fractional Variation operators of order* $0 \leq \beta \leq 1$ *as*

$$v_{\epsilon\pm}^{\beta}\left[f\right](x) := \frac{\Delta_{\epsilon}^{\pm}\left[f\right](x)}{\epsilon^{\beta}} \tag{1}$$

for a positive ϵ.

Definition 2 (Fractional order velocity). *Define the* **fractional velocity** *of fractional order* β *as the limit*

$$v_{\pm}^{\beta} f(x) := \lim_{\epsilon \to 0} \frac{\Delta_{\epsilon}^{\pm}[f](x)}{\epsilon^{\beta}} = \lim_{\epsilon \to 0} v_{\epsilon\pm}^{\beta}\left[f\right](x) \ . \tag{2}$$

A function for which at least one of $v_{\pm}^{\beta} f(x)$ *exists finitely will be called* β*-differentiable at the point* x.

The terms β-velocity and fractional velocity will be used interchangeably throughout the paper. In the above definition we do not require upfront equality of left and right β-velocities. This amounts to not demanding continuity of the β-velocities in advance. Instead, continuity is a property, which is fulfilled under certain conditions. It was further established that for fractional orders fractional velocity is continuous only if it is zero [24].

Further, the following technical conditions are important for applications.

Condition 1 (Hölder growth condition). *For given* x *and* $0 < \beta \leq 1$

$$\operatorname{osc}_{\epsilon}^{\pm} f(x) \leq C\epsilon^{\beta} \tag{C1}$$

for some $C \geq 0$ *and* $\epsilon > 0$.

Condition 2 (Hölder oscillation condition). *For given* x, $0 < \beta \leq 1$ *and* $\epsilon > 0$

$$\operatorname{osc}^{\pm} v_{\epsilon\pm}^{\beta} [f](x) = 0\,. \tag{C2}$$

The conditions for the existence of the fractional velocity were demonstrated in [24]. The main result is repeated here for convenience.

Theorem 1 (Conditions for existence of β-velocity). *For each* $\beta > 0$ *if* $v_{+}^{\beta} f(x)$ *exists (finitely), then* f *is right-Hölder continuous of order* β *at* x *and C1 holds, and the analogous result holds for* $v_{-}^{\beta} f(x)$ *and left-Hölder continuity.*

Conversely, if C2 holds then $v_{\pm}^{\beta} f(x)$ *exists finitely. Moreover, C2 implies C1.*

The proof is given in [24]. The essential algebraic properties of the fractional velocity are given in Appendix B. Fractional velocities provide a local way of approximating the growth of Hölder functions in the following way.

Proposition 1 (Fractional Taylor-Lagrange property). *The existence of* $v_{\pm}^{\beta} f(x) \neq 0$ *for* $\beta \leq 1$ *implies that*

$$f(x \pm \epsilon) = f(x) \pm v_{\pm}^{\beta} f(x)\, \epsilon^{\beta} + \mathcal{O}\left(\epsilon^{\beta}\right)\,. \tag{3}$$

While if

$$f(x \pm \epsilon) = f(x) \pm K\epsilon^{\beta} + \gamma_{\epsilon}\, \epsilon^{\beta}$$

uniformly in the interval $x \in [x, x+\epsilon]$ *for some Cauchy sequence* $\gamma_{\epsilon} = \mathcal{O}(1)$ *and* $K \neq 0$ *is constant in* ϵ *then* $v_{\pm}^{\beta} f(x) = K$.

The proof is given in [24].

Remark 1. *The fractional Taylor-Lagrange property was assumed and applied to establish a fractional conservation of mass formula in ([4], Section 4) assuming the existence of a fractional Taylor expansion according to Odibat and Shawagfeh [25]. These authors derived fractional Taylor series development using repeated application of Caputo's fractional derivative [25].*

The Hölder growth property can be generalized further into the concept of F-analytic functions (see Appendix A for definition). An F-analytic function can be characterized up to the leading fractional order in terms of its α-differentiability.

Proposition 2. *Suppose that* $f \in \mathbb{F}^{E}$ *in the interval* $I = [x, x+\epsilon]$. *Then* $v_{\pm}^{\alpha} f(x)$ *exists finitely for* $\alpha \in [0, \min \mathbb{E}]$.

Proof. The proof follows directly from Proposition 1 observing that

$$\sum_{\alpha_j \in E \setminus \{\alpha_1\}} a_j (x - b_j) = \mathcal{O}\left(|x - b_j|^{\alpha_1}\right)$$

so that using the notation in Definition A4

$$v_{+}^{\alpha_1} f(x) = a_1, \quad v_{-}^{\alpha_1} f(x) = 0$$

□

Remark 2. *From the proof of the proposition one can also see the fundamental asymmetry between the forward and backward fractional velocities. A way to combine this is to define a complex mapping*

$$v_C^{\beta} f(x) := v_+^{\beta} f(x) + v_-^{\beta} f(x) \pm i \left(v_+^{\beta} f(x) - v_-^{\beta} f(x) \right)$$

which is related to the approach taken by Nottale [17] by using complexified velocity operators. However, such unified treatment will not be pursued in this work.

3. Characterization of Singular Functions

3.1. Scale Embedding of Fractional Velocities

As demonstrated previously, the fractional velocity has only "few" non-zero values [24,26]. Therefore, it is useful to discriminate the set of arguments where the fractional velocity does not vanish.

Definition 3. *The set of points where the fractional velocity exists finitely and $v_{\pm}^{\beta} f(x) \neq 0$ will be denoted as the* ***set of change*** $\chi_{\pm}^{\beta}(f) := \left\{ x : v_{\pm}^{\beta} f(x) \neq 0 \right\}$.

Since the set of change $\chi_{+}^{\alpha}(f)$ is totally disconnected [24] some of the useful properties of ordinary derivatives, notably the continuity and the semi-group composition property, are lost. Therefore, if we wish to retain the continuity of description we need to pursue a different approach, which should be equivalent in limit. Moreover, we can treat the fractional order of differentiability (which coincides with the point-wise Hölder exponent, that is Condition C1) as a parameter to be determined from the functional expression.

One can define two types of **scale-dependent operators** for a wide variety if physical signals. An extreme case of such signals are the singular functions

Since for a singular signal the derivative either vanishes or it diverges then the rate of change for such signals cannot be characterized in terms of derivatives. One could apply to such signals either the *fractal variation operators* of certain order or the difference quotients as Nottale does and avoid taking the limit. Alternately, as will be demonstrated further, the scale embedding approach can be used to reach the same goal.

Singular functions can arise as point-wise limits of continuously differentiable ones. Since the fractional velocity of a continuously-differentiable function vanishes we are lead to postpone taking the limit and only apply L'Hôpital's rule, which under the hypotheses detailed further will reach the same limit as $\epsilon \to 0$. Therefore, we are set to introduce another pair of operators which in limit are equivalent to the fractional velocities notably these are the left (resp. right) *scale velocity* operators:

$$\mathcal{S}_{\epsilon\pm}^{\beta} [f](x) := \frac{\epsilon^{\beta}}{\{\beta\}_1} \frac{\partial}{\partial \epsilon} f(x \pm \epsilon) \tag{4}$$

where $\{\beta\}_1 \equiv 1 - \beta \bmod 1$. The ϵ parameter, which is not necessarily small, represents the scale of observation.

The equivalence in limit is guaranteed by the following result:

Proposition 3. *Let $f'(x)$ be continuous and non-vanishing in $I = (x, x \pm \mu)$. That is $f \in AC[I]$. Then*

$$\lim_{\epsilon \to 0} v_{\epsilon\pm}^{1-\beta} [f](x) = \lim_{\epsilon \to 0} \mathcal{S}_{\epsilon\pm}^{\beta} [f](x)$$

if one of the limits exits.

The proof is given in [20] and will not be repeated. In this formulation the value of $1 - \beta$ can be considered as the magnitude of deviation from linearity (or differentiability) of the signal at that point.

Theorem 2 (Scale velocity fixed point). *Suppose that $f \in BVC[x, x+\epsilon]$ and f' does not vanish a.e. in $[x, x+\epsilon]$. Suppose that $\phi \in \mathbb{C}^1$ is a contraction map. Let $f_n(x) := \underbrace{\phi \circ \ldots \phi}_{n} \circ f(x)$ be the n-fold composition and $F(x) := \lim_{n \to \infty} f_n(x)$ exists finitely. Then the following commuting diagram holds:*

$$
\begin{array}{ccc}
f_n(x) & \xrightarrow{\mathcal{S}_{\epsilon\pm}^{1-\beta}} & \mathcal{S}_{\epsilon\pm}^{1-\beta}\left[f_n\right](x) \\
\Big\downarrow \lim_{n\to\infty} & & \epsilon_n \Big\downarrow \lim_{n\to\infty} \\
F(x) & \xrightarrow{v_{\pm}^{\beta}} & v_{\pm}^{\beta} F(x)
\end{array}
$$

The limit in n is taken point-wise.

Proof. The proof follows by induction. Only the forward case will be proven. The backward case follows by reflection of the function argument. Consider an arbitrary n and an interval $I = [x,\, x+\epsilon]$. By differentiability of the map ϕ

$$v_{+}^{\beta} f_n(x) = \left(\frac{\partial \phi}{\partial f}\right)^n v_{+}^{\beta} f(x)$$

Then by hypothesis $f'(x)$ exists finitely a.e. in I so that

$$v_{+}^{\beta} f(x) = \frac{1}{\beta} \lim_{\epsilon \to 0} \epsilon^{1-\beta} f'(x+\epsilon)$$

so that

$$v_{+}^{\beta} f_n(x) = \frac{1}{\beta} \left(\frac{\partial \phi}{\partial f}\right)^n \lim_{\epsilon \to 0} \epsilon^{1-\beta} f'(x+\epsilon)$$

On the other hand

$$\mathcal{S}_{\epsilon+}^{1-\beta}\left[f_n\right](x) = \frac{1}{\beta} \left(\frac{\partial \phi}{\partial f}\right)^n \epsilon^{1-\beta} f'(x+\epsilon)$$

Therefore, if the RHS limits exist they are the same. Therefore, the equality is asserted for all n. Suppose that $v_{+}^{\beta} F(x)$ exists finitely.

Since ϕ is a contraction and $F(x)$ is its fixed point then by Banach fixed point theorem there is a Lipschitz constant $q < 1$ such that

$$\underbrace{|F(x+\epsilon) - f_n(x+\epsilon)|}_{A} \le \frac{q^n}{1-q} |\phi \circ f(x+\epsilon) - f(x+\epsilon)|$$

$$\underbrace{|F(x) - f_n(x)|}_{B} \le \frac{q^n}{1-q} |\phi \circ f(x) - f(x)|$$

Then by the triangle inequality

$$|\Delta_{\epsilon}^{+}[F](x) - \Delta_{\epsilon}^{+}[f_n](x)| \le A + B \le \frac{q^n}{1-q} \left(|\phi \circ f(x+\epsilon) - f(x+\epsilon)| + |\phi \circ f(x) - f(x)|\right) \le q^n L$$

for some L. Then

$$|v_{\epsilon+}^{\beta}[F](x) - v_{\epsilon+}^{\beta}[f_n](x)| \le \frac{q^n L}{\epsilon^{\beta}}$$

We evaluate $\epsilon = q^n/\lambda$ for some $\lambda \ge 1$ so that

$$|v_{\epsilon+}^{\beta}[F](x) - v_{\epsilon+}^{\beta}[f_n](x)| \le q^{n(1-\beta)} L \lambda^{\beta} =: \mu_n$$

Therefore, in limit RHS $\lim_{n\to\infty} \mu_n = 0$. Therefore,

$$\lim_{n\to\infty} v^{\beta}_{\epsilon+} [f_n] (x) = v^{\beta}_{+} F (x) \quad | \quad \epsilon_n \to 0$$

Therefore, by continuity of μ_n in the λ variable the claim follows. □

So stated, the theorem holds also for sets of maps ϕ_k acting in sequence as they can be identified with an action of a single map.

Corollary 1. *Let $\Phi = \{\phi_k\}$, where the domains of ϕ_k are disjoint and the hypotheses of Theorem 2 hold. Then Theorem 2 holds for Φ.*

Corollary 2. *Under the hypotheses of Theorem 2 for $f \in \mathbb{H}^{\alpha}$ and $\frac{\partial\phi}{\partial f} > 1$ there is a Cauchy null sequence $\{\epsilon\}_k^{\infty}$, such that*

$$\lim_{n\to\infty} \mathcal{S}^{1-\alpha}_{\epsilon_n\pm} f_n(x) = v^{\alpha}_{\pm} f(x)$$

This sequence will be named ***scale–regularizing*** *sequence.*

Proof. In the proof of the theorem it was established that

$$\mathcal{S}^{1-\alpha}_{\epsilon+} [f_n] (x) = \frac{1}{\beta} \left(\frac{\partial\phi}{\partial f} \right)^{n} \epsilon^{1-\alpha} f'(x+\epsilon)$$

Then we can identify

$$\epsilon_n = \pm\epsilon \Bigg/ \left| \frac{\partial\phi}{\partial f} \right|^{n/(1-\alpha)}$$

for some $\epsilon < 1$ so that

$$\mathcal{S}^{1-\alpha}_{\epsilon+} [f_n] (x) = \mathcal{S}^{1-\alpha}_{\epsilon_n} [f](x)$$

Then since $\frac{\partial\phi}{\partial f} > 1$ we have $\epsilon_{n+1} < \epsilon_n < 1$. Therefore, $\{\epsilon_k\}_k$ is a Cauchy sequence. Further, the RHS limit evaluates to (omitting n for simplicity)

$$\lim_{\epsilon\to 0} \mathcal{S}^{1-\alpha}_{\epsilon\pm} [f] (x) = \frac{1}{\alpha} \lim_{\epsilon\to 0} \epsilon^{1-\alpha} f'(x\pm\epsilon) = v^{\alpha}_{\pm} f(x)$$

The backward case can be proven by identical reasoning. □

Therefore, we can identify the Lipschitz constant q by the properties of the contraction maps as will be demonstrated in the following examples.

3.2. Applications

3.2.1. De Rham Function

De Rham's function arises in several applications. Lomnicki and Ulan [27] have given a probabilistic construction. In a an imaginary experiment of flipping a possibly "unfair" coin with probability a of heads (and $1-a$ of tails). Then $R_a(x) = \mathbb{P}\{t \le x\}$ after infinitely many trials where t is the record of the trials.

Mandelbrot [13] introduces a multiplicative cascade, describing energy dissipation in turbulance, which is related to the increments of the De Rham's function.

The function can be defined in different ways [22,28,29]. One way to define the De Rham's function is as the unique solutions of the functional equations

$$R_a(x) := \begin{cases} aR_a(2x), & 0 \le x < \frac{1}{2} \\ (1-a)R_a(2x-1)+a, & \frac{1}{2} \le x \le 1 \end{cases}$$

and boundary values $R_a(0) = 0$, $R_a(1) = 1$. The function is strictly increasing and singular. Its scaling properties and additional functional equations are described in [30].

In another re-parametrization the defining functional equations become

$$R_a(x) = aR_a(2x)$$
$$R_a(x+1/2) = (1-a)R_a(2x) + a$$

In addition, there is a symmetry with respect to inversion about 1.

$$R_a(x) = 1 - R_{1-a}(1-x)$$

De Rham's function is also known under several different names—"Lebesgue's singular function" or "Salem's singular function".

De Rham's function can be re-parametrized on the basis of the point-wise Hölder exponent [20]. Then it is the fixed point of the following IFS:

$$r_n(x,\alpha) := \begin{cases} x^\alpha, & n = 0 \\ \frac{1}{2^\alpha} r_{n-1}(2x,\alpha), & 0 \le x < \frac{1}{2} \\ (1-\frac{1}{2^\alpha}) r_{n-1}(2x-1,\alpha) + \frac{1}{2^\alpha}, & \frac{1}{2} \le x \le 1 \end{cases}$$

provided that $a = 1/2^\alpha, a \ge 1/2$. For the case $a \le 1/2$ the parametrization corresponding to the original IFS is $\alpha = -\log_2(1-a)$. The IFS converges point-wise to $R_\alpha(x) = \lim_{n\to\infty} r_n(x,\alpha)$.

Formal calculation shows that

$$\upsilon_+^\beta r_n(x,\alpha) := \begin{cases} 2^{\beta-\alpha} \upsilon_+^\beta r_{n-1}(2x,\alpha), & 0 \le x < \frac{1}{2} \\ (2^\alpha - 1) 2^{\beta-\alpha} \upsilon_+^\beta r_{n-1}(2x-1,\alpha), & \frac{1}{2} \le x \le 1 \end{cases}$$

Therefore, for $\beta < \alpha$ the fractional velocity vanishes, while for $\beta > \alpha$ it diverges. We further demonstrate its existence for $\beta = \alpha$. For this case formally

$$\upsilon_+^\alpha r_n(x,\alpha) := \begin{cases} \upsilon_+^\alpha r_{n-1}(2x,\alpha), & 0 \le x < \frac{1}{2} \\ (2^\alpha - 1) \upsilon_+^\alpha r_{n-1}(2x-1,\alpha), & \frac{1}{2} \le x \le 1 \end{cases}$$

and $\upsilon_+^a r_0(x,\alpha) = \mathbf{1}(x=0)$.

The same result can be reached using scaling arguments. We can discern two cases.

Case 1, $x \le 1/2$: Then application of the scale operator leads to :

$$\mathcal{S}_{\epsilon+}^{1-\alpha}[r_n](x,\alpha) = \frac{2^{1-\alpha}\epsilon^{1-\alpha}}{\alpha} \frac{\partial}{\partial\epsilon} r_{n-1}(2x+2\epsilon,\alpha)$$

Therefore, the pre-factor will remain scale invariant for $\epsilon = 1/2$ and consecutively $\epsilon_n = \frac{1}{2^n}$ so that we identify a scale-regularizing Cauchy sequence so that $f(x) = x^a$ is verified and $\upsilon_+^\alpha R_a(0) = 1$.

Case 2, $x > 1/2$: In a similar way :

$$\mathcal{S}_{\epsilon+}^{1-\alpha}[r_n](x,\alpha) = \frac{2^{1-\alpha}\epsilon^{1-\alpha}}{\alpha} (2^\alpha - 1) \frac{\partial}{\partial\epsilon} r_{n-1}(2x+2\epsilon-1,\alpha)$$

Applying the same sequence results in a factor $2^a - 1 \le 1$. Therefore, the resulting transformation is a contraction.

Finally, we observe that if $x = \overline{0.d_1 \dots d_n}$ is in binary representation then the number of occurrences of Case 2 corresponds to the number of occurrences of the digit 1 in the binary notation (see below).

The calculation can be summarized in the following proposition:

Proposition 4. *Let $\mathbb{Q}_2$ denote the set of dyadic rationals. Let $s_n = \sum_{k=1}^{n} d_k$ denote the sum of the digits for the number $x = \overline{0.d_1 \dots d_n}$, $d \in \{0,1\}$ in binary representation, then*

$$v_+^{\alpha} R_a(x) = \begin{cases} \left(2^{\beta} - 1\right)^{s_n}, & x \in \mathbb{Q}_2 \\ 0, & x \notin \mathbb{Q}_2 \end{cases}$$

for $\alpha = -log_2 a$, $a > \frac{1}{2}$. For $\alpha < -log_2 a$ $v_+^{\alpha} R_a(x) = 0$.

3.2.2. Bernoulli-Mandelbrot Binomial Measure

The binomial Mandlebrot measure m_a is constructed as follows. Let $\mathcal{T}_n$ be the complete partition of dyadic intervals of the n-th generation on $\mathcal{T}_0 := [0,1)$. In addition, let's assume the generative scheme

$$\mathcal{T}_n \mapsto (\mathcal{T}_{n+1}^L, \mathcal{T}_{n+1}^R)$$

where the interval $\mathcal{T}_n = \left[x_n,\ x_n + \frac{1}{2^n}\right)$ is split into a left and right children such that

$$\mathcal{T}_{n+1}^L \bigcup \mathcal{T}_{n+1}^R = \mathcal{T}_n,\ \mathcal{T}_{n+1}^L \bigcap \mathcal{T}_{n+1}^R = \emptyset.$$

Let

$$\mathcal{T}_{n+1}^L = \left[x_n^L,\ x_{n+1}^L + \frac{1}{2^{n+1}}\right), \quad \mathcal{T}_{n+1}^R = \left[x_n^R,\ x_{n+1}^R + \frac{1}{2^{n+1}}\right).$$

Then define the measure m_a recursively as follows:

$$\begin{aligned} m_a(\mathcal{T}_{n+1}^L) &= a\, m_a(\mathcal{T}_n), \\ m_a(\mathcal{T}_{n+1}^R) &= (1-a)\, m_a(\mathcal{T}_n), \\ m_a(\mathcal{T}_0) &= 1. \end{aligned}$$

The construction is presented in the diagram below:

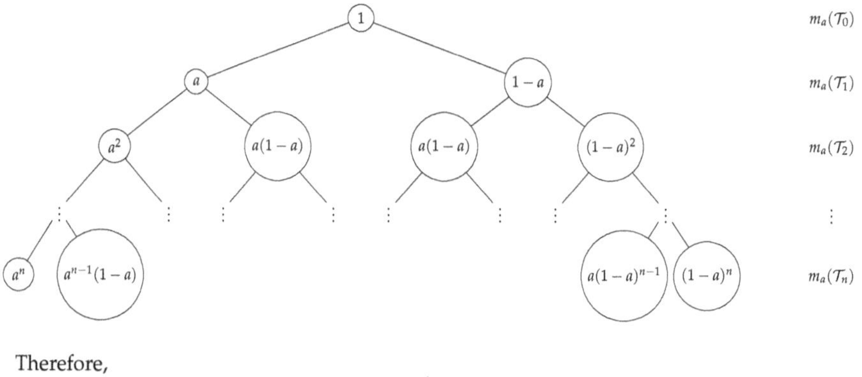

Therefore,

$$x_{n+1}^L + \frac{1}{2^{n+1}} = x_{n+1}^R$$

so that all $x_n \in \mathbb{Q}_2$.

To elucidate the link to the Mandlebrot measure we turn to the arithmetic representation of the De Rham function. That is, consider $x = \overline{0.d_1 \ldots d_n}$, $d \in \{0,1\}$ under the convention that for a dyadic rational the representation terminates. Then according to Lomnicki and Ulam [27]

$$R_a(x) = \sum_{k=1}^{n} d_k a^{k-s_n+1}(1-a)^{s_n-1}\,, s_n = \sum_{k=1}^{n} d_k$$

Consider the increment of $R_a(x)$ for $\epsilon = 1/2^{n+1}$. If $x = \overline{0.d_1 \ldots d_n}$ then $x + 1/2^{n+1} = \overline{0.d_1 \ldots d_n 1}$ so that

$$D(x) := R_a\left(x + 1/2^{n+1}\right) - R_a\left(x\right) = a^{n+1-s_n}(1-a)^{s_n} = a^{n+1}\left(1/a - 1\right)^{s_n}.$$

From this it is apparent that

$$D(x) = aD(2x), x < 1/2 \quad D(x) = (1-a)D(2x-1), x \geq 1/2$$

Therefore, we can identify $D(x) = m_a(\mathcal{T})$ for $x \in \mathcal{T}$.
On the other hand, for $a = 1/2^{\alpha}$

$$\upsilon^{\alpha}_{\epsilon+}\left[R_a\right](x) = \frac{D(x)}{1/2^{(n+1)\alpha}} = 2^{(n+1)\alpha} m_a(\mathcal{T}_{n+1})$$

for $x \in \mathcal{T}_{n+1}$.

Remark 3. *It should be noted that based on the symmetry equation*

$$\upsilon^{\alpha}_{+} R_a\left(x\right) = \upsilon^{\alpha}_{-} R_a\left(1-x\right)$$

so that $\upsilon^{\alpha}_{+} R_a\left(x\right) \neq \upsilon^{\alpha}_{-} R_a\left(x\right)$ *for* $\alpha \neq 1$ *so that the measure* m_a *does not have a density in agreement with Marstrand's theorem.*

3.2.3. Neidinger Function

Neidinger introduces a novel strictly singular function [31], which he called *fair-bold gambling function*. The function is based on De Rham's construction. The Neidinger's function is defined as the limit of the system

$$N_n(x,a) := \begin{cases} x, & n = 0 \\ a \leftarrow 1-a & \\ aN_{n-1}(2x, 1-a), & 0 \leq x < \frac{1}{2} \\ (1-a)\,N_{n-1}(2x-1, 1-a) + a, & \frac{1}{2} \leq x \leq 1 \end{cases}$$

In other words the parameter a alternates for every recursion step but is not passed on the next function call (see Figure 1).

We can exercise a similar calculation again starting from $r_0(x,a) = x^a$. Then

$$\upsilon^{\beta}_{+} r_n(x,a) := \begin{cases} a \leftarrow 1-a, & n \quad even \\ a2^{\beta}\upsilon^{\beta}_{+} r_{n-1}(2x, a), & 0 \leq x < \frac{1}{2} \\ (1-a)2^{\beta}\upsilon^{\beta}_{+} r_{n-1}(2x-1, a), & \frac{1}{2} \leq x \leq 1 \end{cases}$$

Therefore, either $a = 1/2^{\beta}$ or $1 - a = 1/2^{\beta}$ so that $\upsilon^{a}_{+} r_0(x,a) = \mathbf{1}(x=0)$. The velocity can be computed algorithmically to arbitrary precision (Figure 2).

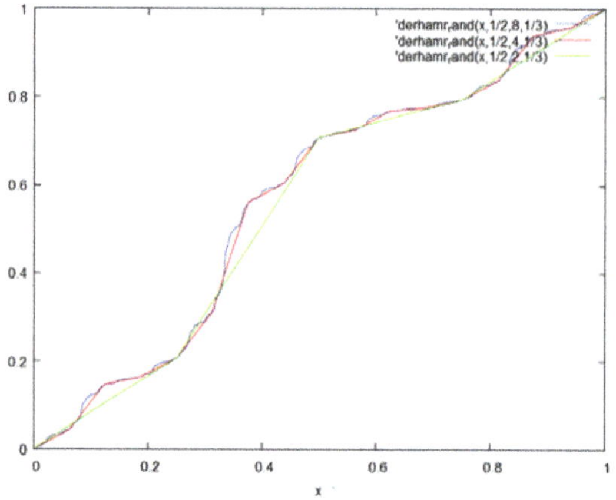

Figure 1. Recursive construction of the Neidinger's function; iteration levels 2, 4, 8.

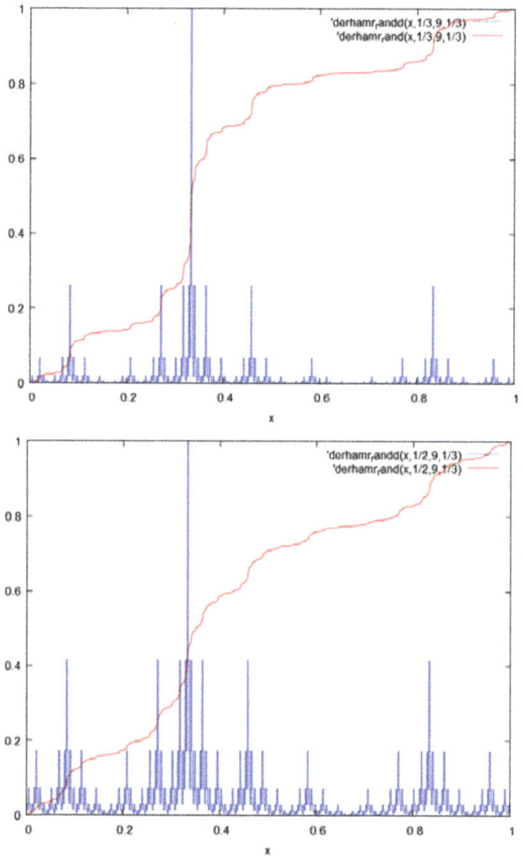

Figure 2. Approximation of the fractional velocity of Neidinger's function. Recursive construction of the fractional velocity for $\beta = 1/3$ (**top**) and $\beta = 1/2$ (**bottom**), iteration level 9. The Neidinger's function IFS are given for comparison for the same iteration level.

3.2.4. Langevin Evolution

Consider a non-linear problem, where the continuous phase-space trajectory of a system is represented by a F-analytic function $x(t)$ and t is a real-valued parameter, for example time or curve length. That is, suppose that a generalized Langevin equation holds uniformly in $[t, t+\epsilon]$ for measurable functions a,B :

$$\Delta_{\epsilon}^{+}[x](t) = a(x,t)\epsilon + B(x,t)\epsilon^{\beta} + \mathcal{O}(\epsilon),\ \beta \leq 1$$

The form of the equation depends critically on the assumption of continuity of the reconstructed trajectory. This in turn demands for the fluctuations of the fractional term to be discontinuous. The proof technique is introduced in [24], while the argument is similar to the one presented in [32].

By hypothesis $\exists K$, such that $|\Delta_{\epsilon}X| \leq K\epsilon^{\beta}$ and $x(t)$ is $\mathbb{H}^{\beta}$. Therefore, without loss of generality we can set $a = 0$ and apply the argument from [24]. Fix the interval $[t, t+\epsilon]$ and choose a partition of points $\{t_k = t + k/N\epsilon\}$ for an integral N.

$$x_{t_k} = x_{t_{k-1}} + B(x_{t_{k-1}}, t_{k-1})\,(\epsilon/N)^{\beta} + \mathcal{O}\left(\epsilon^{\beta}\right)$$

where we have set $x_{t_k} \equiv x(t_k)$. Therefore,

$$\Delta_{\epsilon}x = \frac{1}{N^{\beta}}\sum_{k=0}^{N-1} B(x_{t_k}, t_k)\epsilon^{\beta} + \mathcal{O}\left(\epsilon^{\beta}\right)$$

Therefore, if we suppose that B is continuous in x or t after taking limit on both sides we arrive at

$$\limsup_{\epsilon\to 0}\frac{\Delta_{\epsilon}x}{\epsilon^{\beta}} = B(x_t, t) = \frac{1}{N^{\beta}}\sum_{k=0}^{N-1}\limsup_{\epsilon\to 0} B(x_{t_k}, t_k) = N^{1-\beta}B(x_t, t)$$

so that $(1 - N^{1-\beta})B(x,t) = 0$. Therefore, either $\beta = 1$ or $B(x,t) = 0$. So that $B(x,t)$ must oscillate and is not continuous if $\beta < 1$.

3.2.5. Brownian Motion

The presentation of this example follows [33]. However, in contrast to Zili $\upsilon_{+}^{\alpha} = \upsilon_{-}^{\alpha}$ is not assumed point-wise. Consider the Brownian motion W_t. Using the stationarity and self-similarity of the increments $\Delta_{\epsilon}^{+}W_t = \sqrt{\epsilon}\,N(0,1)$ where $N(0,1)$ is a Gaussian random variable. Therefore,

$$\upsilon_{\pm}^{\alpha}W_t = 0, \quad \alpha < 1/2$$

while $\upsilon_{\pm}^{\alpha}W_t$ does not exist for $\alpha > 1/2$ with probability $\mathbb{P} = 1$. The estimate holds a.s. since $\mathbb{P}(\Delta_{\epsilon}^{+}W_t = 0) = 0$. Interestingly, for $\alpha = 1/2$ the velocity can be regularized to a finite value if we take the expectation. That is

$$\upsilon_{+}^{\alpha}\,\mathbb{E}W_t = 0$$

since $\Delta_{\epsilon}^{+}\mathbb{E}W_t = 0$. Also $\upsilon_{+}^{\alpha}\,\mathbb{E}|W_t| = 1$.

4. Characterization of Kolwankar-Gangal Local Fractional Derivatives

The overlap of the definitions of the Cherebit's fractional velocity and the Kolwankar-Gangal fractional derivative is not complete [34]. Notably, Kolwankar-Gangal fractional derivatives are sensitive to the critical local Hölder exponents, while the fractional velocities are sensitive to the critical point-wise Hölder exponents and there is no complete equivalence between those quantities [35]. In this section we will characterize the local fractional derivatives in the sense of Kolwankar and Gangal using the notion of fractional velocity.

4.1. Fractional Integrals and Derivatives

The left Riemann-Liouville differ-integral of order $\beta \geq 0$ is defined as

$$_{a+}\mathbf{I}_x^{\beta} f(x) = \frac{1}{\Gamma(\beta)} \int_a^x f(t)\,(x-t)^{\beta-1}\,dt$$

while the right integral is defined as

$$_{-a}\mathbf{I}_x^{\beta} f(x) = \frac{1}{\Gamma(\beta)} \int_x^a f(t)\,(t-x)^{\beta-1}\,dt$$

where $\Gamma(x)$ is the Euler's Gamma function (Samko et al. [36], p. 33). The left (resp. right) Riemann-Liouville (R-L) fractional derivatives are defined as the expressions (Samko et al. [36], p. 35):

$$\mathcal{D}^{\beta}_{a+} f(x) := \frac{d}{dx}\, {}_{a+}\mathbf{I}_x^{1-\beta} f(x) = \frac{1}{\Gamma(1-\beta)} \frac{d}{dx} \int_a^x \frac{f(t)}{(x-t)^{\beta}} dt$$
$$\mathcal{D}^{\beta}_{-a} f(x) := -\frac{d}{dx}\, {}_{-a}\mathbf{I}_x^{1-\beta} f(x) = -\frac{1}{\Gamma(1-\beta)} \frac{d}{dx} \int_x^a \frac{f(t)}{(t-x)^{\beta}} dt$$

The left (resp. right) R-L derivative of a function f exists for functions representable by fractional integrals of order α of some Lebesgue-integrable function. This is the spirit of the definition of Samko et al. ([36], Definition 2.3, p. 43) for the appropriate functional spaces:

$$\mathcal{I}^{\alpha}_{a,+}(L^1) := \left\{ f : {}_{a+}\mathbf{I}_x^{\alpha} f(x) \in AC([a,b]), f \in L^1([a,b]), x \in [a,b] \right\},$$
$$\mathcal{I}^{\alpha}_{a,-}(L^1) := \left\{ f : {}_{-a}\mathbf{I}_x^{\alpha} f(x) \in AC([a,b]), f \in L^1([a,b]), x \in [a,b] \right\}$$

Here AC denotes absolute continuity on an interval in the conventional sense. Samko et al. comment that the existence of a summable derivative $f'(x)$ of a function $f(x)$ does not yet guarantee the restoration of $f(x)$ by the primitive in the sense of integration and go on to give references to singular functions for which the derivative vanishes almost everywhere and yet the function is not constant, such as for example, the De Rhams's function [37].

To ensure restoration of the primitive by fractional integration, based on Th. 2.3 Samko et al. introduce another space of *summable fractional derivatives*, for which the Fundamental Theorem of Fractional Calculus holds.

Definition 4. *Define the functional spaces of summable fractional derivatives Samko et al. ([36], Definition 2.4, p. 44) as* $E^{\alpha}_{a,\pm}([a,b]) := \left\{ f : \mathcal{I}^{1-\alpha}_{a,\pm}(L^1) \right\}$.

In this sense

$$_{a+}\mathbf{I}_x^{\alpha} \left(\mathcal{D}^{\alpha}_{a+} f \right)(x) = f(x)$$

for $f \in E^{\alpha}_{a,+}([a,b])$ (Samko et al. [36], Theorem 4, p. 44). While

$$\mathcal{D}^{\alpha}_{a+} \left({}_{a+}\mathbf{I}_x^{\alpha} f \right)(x) = f(x)$$

for $f \in \mathcal{I}^{\alpha}_{a,+}(L^1)$.

So defined spaces do not coincide. The distinction can be seen from the following example:

Example 1. *Define*

$$h(x) := \begin{cases} 0, & x \leq 0 \\ x^{\alpha-1}, & x > 0 \end{cases}$$

for $0 < \alpha < 1$. *Then* ${}_{0+}\mathbf{I}_x^{1-\alpha} h(x) = \Gamma(\alpha)$ *for* $x > 0$ *so that* $\mathcal{D}_{0+}^{\alpha}\, h(x) = 0$ *everywhere in* $\mathbb{R}$.

On the other hand,

$$ {}_{0+}\mathbf{I}_x^{\alpha} h(x) = \frac{\Gamma(\alpha)}{\Gamma(2\alpha)}\, x^{2\alpha-1} $$

for $x > 0$ *and*

$$ \frac{\Gamma(\alpha)}{\Gamma(2\alpha)}\, \mathcal{D}_{0+}^{\alpha} x^{2\alpha-1} = x^{\alpha-1} \,. $$

Therefore, the fundamental theorem fails. It is easy to demonstrate that $h(x)$ *is not AC on any interval involving* 0.

Therefore, the there is an inclusion $E_{a,+}^{\alpha} \subset \mathcal{I}_{a,+}^{\alpha}$.

4.2. The Local(ized) Fractional Derivative

The definition of *local fractional derivative* (LFD) introduced by Kolwankar and Gangal [38] is based on the localization of Riemann-Liouville fractional derivatives towards a particular point of interest in a way similar to Caputo.

Definition 5. *Define left LFD as*

$$ \mathcal{D}_{KG+}^{\beta} f(x) := \lim_{x \to a} \mathcal{D}_{a+}^{\beta} \left[f - f(a) \right](x) $$

and right LFD as

$$ \mathcal{D}_{KG-}^{\beta} f(x) := \lim_{x \to a} \mathcal{D}_{-a}^{\beta} \left[f(a) - f \right](x) \,. $$

Remark 4. *The seminal publication defined only the left derivative. Note that the LFD is more restrictive than the R-L derivative because the latter may not have a limit as* $x \to a$.

Ben Adda and Cresson [23] and later Chen et al. [26] claimed that the Kolwankar—Gangal definition of local fractional derivative is equivalent to Cherbit's definition for certain classes of functions. On the other hand, some inaccuracies can be identified in these articles [26,34]. Since the results of Chen et al. [26] and Ben Adda-Cresson [34] are proven under different hypotheses and notations I feel that separate proofs of the equivalence results using the theory established so-far are in order.

Proposition 5 (LFD equivalence). *Let* $f(x)$ *be* β*-differentiable about* x. *Then* $\mathcal{D}_{KG,\pm}^{\beta} f(x)$ *exists and*

$$ \mathcal{D}_{KG,\pm}^{\beta} f(x) = \Gamma(1+\beta)\, v_{\pm}^{\beta} f(x) \,. $$

Proof. We will assume that $f(x)$ is non-decreasing in the interval $[a, a+x]$. Since x will vary, for simplicity let's assume that $v_{+}^{\beta} f(a) \in \chi^{\beta}$. Then by Proposition 1 we have

$$ f(z) = f(a) + v_{+}^{\beta} f(a)\,(z-a)^{\beta} + o\left((z-a)^{\beta}\right), \; a \le z \le x \,. $$

Standard treatments of the fractional derivatives [8] and the changes of variables $u = (t-a)/(x-a)$ give the alternative Euler integral formulation

$$ \mathcal{D}_{+a}^{\beta} f(x) = \frac{\partial}{\partial h} \left(\frac{h^{1-\beta}}{\Gamma(1-\beta)} \int_0^1 \frac{f(hu+a) - f(a)}{(1-u)^{\beta}} du \right) \tag{5} $$

for $h = x - a$. Therefore, we can evaluate the fractional Riemann-Liouville integral as follows:

$$\frac{h^{1-\beta}}{\Gamma(1-\beta)}\int\limits_0^1 \frac{f(hu+a)-f(a)}{(1-u)^\beta}du = \frac{h^{1-\beta}}{\Gamma(1-\beta)}\int\limits_0^1 \frac{K\,(hu)^\beta + \mathcal{O}\left((hu)^\beta\right)}{(1-u)^\beta}du =: I$$

setting conveniently $K = v_+^\beta f\,(a)$. The last expression I can be evaluated in parts as

$$I = \underbrace{\frac{h^{1-\beta}}{\Gamma(1-\beta)}\int\limits_0^1 \frac{Kh^\beta u^\beta}{(1-u)^\beta}du}_{A} + \underbrace{\frac{h^{1-\beta}}{\Gamma(1-\beta)}\int\limits_0^1 \frac{\mathcal{O}\left((hu)^\beta\right)}{(1-u)^\beta}du}_{C} .$$

The first expression is recognized as the Beta integral [8]:

$$A = \frac{h^{1-\beta}}{\Gamma(1-\beta)}B\,(1-\beta,1+\beta)\,h^\beta K = \Gamma(1+\beta)\,Kh$$

In order to evaluate the second expression we observe that by Proposition 1

$$\left|\mathcal{O}\left((hu)^\beta\right)\right| \le \gamma(hu)^\beta$$

for a positive $\gamma = \mathcal{O}_1$. Assuming without loss of generality that $f(x)$ is non decreasing in the interval we have $C \le \Gamma(1+\beta)\,\gamma h$ and

$$\mathcal{D}_{a+}^\beta f(x) \le (K+\gamma)\,\Gamma(1+\beta)$$

and the limit gives $\lim\limits_{x\to a+} K+\gamma = K$ by the *squeeze lemma* and Proposition 1. Therefore, $\mathcal{D}_{KG+}^\beta f(a) = \Gamma(1+\beta)v_+^\beta f\,(a)$. On the other hand, for $\mathbb{H}^{r,\alpha}$ and $\alpha > \beta$ by the same reasoning

$$A = \frac{h^{1-\beta}}{\Gamma(1-\beta)}B\,(1-\beta,1+\alpha)\,h^\alpha K = \Gamma(1+\beta)\,Kh^{1-\beta+\alpha} .$$

Then differentiation by h gives

$$A'_h = \frac{\Gamma(1+\alpha)}{\Gamma(1+\alpha-\beta)}\,Kh^{\alpha-\beta} .$$

Therefore,

$$\mathcal{D}_{KG+}^\beta f(x) \le \frac{\Gamma(1+\alpha)}{\Gamma(1+\alpha-\beta)}\,(K+\gamma)\,h^{\alpha-\beta}$$

by monotonicity in h. Therefore, $\mathcal{D}_{KG\pm}^\beta f(a) = v_\pm^\beta f\,(a) = 0$. Finally, for $\alpha = 1$ the expression A should be evaluated as the limit $\alpha \to 1$ due to divergence of the Γ function. The proof for the left LFD follows identical reasoning, observing the backward fractional Taylor expansion property. □

Proposition 6. *Suppose that $\mathcal{D}_{KG\pm}^\beta f(x)$ exists finitely and the related R-L derivative is summable in the sense of Definition 4. Then f is β-differentiable about x and $\mathcal{D}_{KG,\pm}^\beta f(x) = \Gamma(1+\beta)\,v_\pm^\beta f\,(x)$.*

Proof. Suppose that $f \in E_{a,+}^\alpha([a, a+\delta])$ and let $\mathcal{D}_{KG+}^\alpha f(x) = L$. The existence of this limit implies the inequality

$$\left|\mathcal{D}_{a+}^\alpha\left[f - f(a)\right](x) - L\right| < \mu$$

for $|x-a| \le \delta$ and a Cauchy sequence μ.

Without loss of generality suppose that $\mathcal{D}^{\alpha}_{a+}\left[f - f(a)\right](x)$ is non-decreasing and $L \neq 0$. We proceed by integrating the inequality:

$$ {}_{a+}\mathbf{I}^{\alpha}_{x} \left(\mathcal{D}^{\alpha}_{a+}\left[f - f(a)\right](x) - L\right) < {}_{a+}\mathbf{I}^{\alpha}_{x}\mu $$

Then by the Fundamental Theorem

$$ f(x) - f(a) - \frac{L}{\Gamma(\alpha)}(x-a)^{\alpha} < \frac{\mu(x-a)^{\alpha}}{\Gamma(\alpha)} $$

and

$$ \frac{f(x) - f(a) - L/\Gamma(\alpha)}{(x-a)^{\alpha}} < \frac{\mu}{\Gamma(\alpha)} = \mathcal{O}(1) $$

which is Cauchy. Therefore, by Proposition 1 f is α-differentiable at x and $\mathcal{D}^{\alpha}_{KG,+}f(x) = \Gamma(1+\alpha)\, v^{\alpha}_{+}f(x)$. The last assertion comes from Proposition 5. The right case can be proven in a similar manner. □

The weaker condition of only point-wise Hölder continuity requires the additional hypothesis of summability as identified in [34]. The following results can be stated.

Lemma 1. *Suppose that $\mathcal{D}^{\beta}_{KG\pm}f(a)$ exists finitely in the weak sense, i.e., implying only that $f \in \mathcal{I}^{\alpha}_{a,+}(L^1)$. Then Condition C1 holds for f a.e. in the interval $[a, x+\epsilon]$.*

Proof. The left R-L derivative can be evaluated as follows. Consider the fractional integral in the Liouville form

$$ \begin{aligned} I_1 &= \int_0^{\epsilon+x-a} \frac{f(x+\epsilon-h) - f(a)}{h^{\beta}}\,dh - \int_0^{x-a} \frac{f(x-h) - f(a)}{h^{\beta}}\,dh \\ &= \underbrace{\int_{x-a}^{\epsilon+x-a} \frac{f(x+\epsilon-h) - f(a)}{h^{\beta}}\,dh}_{I_2} + \underbrace{\int_0^{x-a} \frac{f(x+\epsilon-h) - f(x-h)}{h^{\beta}}\,dh}_{I_3} \end{aligned} $$

Without loss of generality assume that f is non-decreasing in the interval $[a, x+\epsilon-a]$ and set $M_{y,x} = \sup_{[x,y]} f - f(x)$ and $m_{y,x} = \inf_{[x,y]} f - f(x)$. Then

$$ I_2 \leq \int_{x-a}^{\epsilon+x-a} \frac{M_{x+\epsilon,a}}{h^{\beta}}\,dh = \frac{M_{x+\epsilon,a}}{1-\beta}\left[(x-\epsilon+a)^{1-\beta} - (x-a)^{1-\beta}\right] \leq \epsilon \frac{M_{x+\epsilon,a}}{(x-a)^{\beta}} + \mathcal{O}\left(\epsilon^2\right) $$

for $x \neq a$. In a similar manner

$$ I_2 \geq m_{x+\epsilon,a} \frac{\epsilon}{(x-a)^{\beta}} + \mathcal{O}\left(\epsilon^2\right). $$

Then dividing by ϵ gives

$$ \frac{m_{x+\epsilon,a}}{(x-a)^{\beta}} + \mathcal{O}(\epsilon) \leq \frac{I_2}{\epsilon} \leq \frac{M_{x+\epsilon,a}}{(x-a)^{\beta}} + \mathcal{O}(\epsilon) $$

Therefore, the quotient limit is bounded from both sides as

$$ \frac{m_{x,a}}{(x-a)^{\beta}} \leq \underbrace{\lim_{\epsilon\to 0} \frac{I_2}{\epsilon}}_{I_2'} \leq \frac{M_{x,a}}{(x-a)^{\beta}} $$

by the continuity of f. In a similar way we establish

$$I_3 \leq \int_0^{x-a} \frac{M_{x+\epsilon,x}}{h^\beta}\, dh = \frac{M_{x+\epsilon,x}}{1-\beta}\,(x-a)^{1-\beta}$$

and

$$\frac{m_{x+\epsilon,x}}{1-\beta}\,(x-a)^{1-\beta} \leq I_3$$

Therefore,

$$\frac{m_{x+\epsilon,x}}{(1-\beta)\,\epsilon}\,(x-a)^{1-\beta} \leq \frac{I_3}{\epsilon} \leq \frac{M_{x+\epsilon,x}}{(1-\beta)\,\epsilon}\,(x-a)^{1-\beta}$$

By the absolute continuity of the integral the quotient limit $\frac{I_3}{\epsilon}$ exists as $\epsilon \to 0$ for almost all x. This also implies the existence of the other two limits. Therefore, the following bond holds

$$m^\star_{x+\epsilon,x}\frac{(x-a)^{1-\beta}}{(1-\beta)} \leq \underbrace{\lim_{\epsilon\to 0}\frac{I_3}{\epsilon}}_{I_3'} \leq M^\star_{x+\epsilon,x}\frac{(x-a)^{1-\beta}}{(1-\beta)}$$

where $M^\star_{x+\epsilon,x} = \sup_{[x,x+\epsilon]} f'$ and $m^\star_{x+\epsilon,x} = \inf_{[x,x+\epsilon]} f'$ wherever these exist. Therefore, as x approaches a $\lim\limits_{x\to a} I_3' = 0$.

Finally, we establish the bounds of the limit

$$\lim_{x\to a}\frac{m_{x,a}}{(x-a)^\beta} \leq \lim_{x\to a} I_2' \leq \lim_{x\to a}\frac{M_{x,a}}{(x-a)^\beta}\,.$$

Therefore, Condition C1 is necessary for the existing of the limit and hence for $\lim\limits_{x\to a} I'$. □

Based on this result, we can state a generic continuity result for LFD of fractional order.

Theorem 3 (Continuity of LFD). *For $0 < \beta < 1$ if $\mathcal{D}^\beta_{KG\pm} f(x)$ is continuous about x then $\mathcal{D}^\beta_{KG\pm} f(x) = 0$.*

Proof. We will prove the case for $\mathcal{D}^\beta_{KG+} f(x)$. Suppose that LFD is continuous in the interval $[a, x]$ and $\mathcal{D}^\beta_{KG+} f(a) = K \neq 0$. Then the conditions of Lemma 1 apply, that is $f \in \mathbb{H}^\beta$ a.e. in $[a, x]$. Therefore, without loss of generality we can assume that $f \in \mathbb{H}^\beta$ at a. Further, we express the R-L derivative in Euler form setting $z = x - a$:

$$\mathcal{D}^\beta_z f = \frac{\partial}{\partial z}\frac{z^{1-\beta}}{\Gamma(1-\beta)}\int_0^1 \frac{f\left(z\left(1-t\right)+a\right) - f\left(a\right)}{t^\beta}dt$$

By the monotonicity of the power function (e.g., Hölder growth property):

$$k_1\Gamma(1+\beta) \leq \mathcal{D}^\beta_z f \leq K_1\Gamma(1+\beta)$$

where $k_1 = \inf_{[a,a+z]} f - f(a)$ and $K_1 = \sup_{[a,a+z]} f - f(a)$. On the other hand, we can split the integrand in two expressions for an arbitrary intermediate value $z_0 = \lambda z \leq z$. This gives

$$\mathcal{D}^{\beta}_{z} f = \frac{\partial}{\partial z} \frac{z^{1-\beta}}{\Gamma(1-\beta)} \int_0^1 \frac{f\left(z\left(1-t\right)+a\right) - f\left(\lambda z\left(1-t\right)+a\right)}{t^{\beta}} dt +$$

$$\frac{\partial}{\partial z} \frac{z^{1-\beta}}{\Gamma(1-\beta)} \int_0^1 \frac{f\left(\lambda z\left(1-t\right)+a\right) - f\left(a\right)}{t^{\beta}} dt \,.$$

Therefore, by the Hölder growth property and monotonicity in z

$$\mathcal{D}^{\beta}_{z} f \le \frac{\partial}{\partial z} z\left(1-\lambda\right)^{\beta} \Gamma(1+\beta) K_{1-\lambda} + \frac{\partial}{\partial z} z \lambda^{\beta} \Gamma(1+\beta) K_{\lambda} \,.$$

where $K_{\lambda} = \sup_{[a,a+\lambda z]} f - f(a)$ and $K_{1-\lambda} = \sup_{[a+\lambda z, a+z]} f - f(a+\lambda z)$. Therefore,

$$k_1 \Gamma(1+\beta) \le \mathcal{D}^{\beta}_{z} f \le \left((1-\lambda)^{\beta} K_{1-\lambda} + \lambda^{\beta} K_{\lambda} \right) \Gamma(1+\beta) \,.$$

However, by the assumption of continuity $k_1 = K_{\lambda} = K_{1-\lambda} = K$ as $z \to 0$ and the non-strict inequalities become equalities so that

$$\left((1-\lambda)^{\beta} + \lambda^{\beta} - 1 \right) K = 0 \,.$$

However, if $\beta < 1$ we have contradiction since then $\lambda = 1$ or $\lambda = 0$ must hold and λ ceases to be arbitrary. Therefore, since λ is arbitrary $K = 0$ must hold. The right case can be proven in a similar manner. □

Corollary 3 (Discontinuous LFD). *Let* $\chi_{\beta} := \{x : \mathcal{D}^{\beta}_{KG\pm} f(x) \neq 0\}$. *Then for* $0 < \beta < 1$ χ_{β} *is totally disconnected.*

Remark 5. *This result is related to Corollary 3 in [26] however here it is established in a more general way.*

4.3. Equivalent Forms of LFD

LFD can be calculated in the following way. Starting from Formula (5) for convenience we define the integral average

$$M_a(h) := \int_0^1 \frac{f(hu+a) - f(a)}{(1-u)^{\beta}} du \tag{6}$$

Then

$$\Gamma(1-\beta) \mathcal{D}^{\beta}_{KG+} f(a) = \underbrace{\lim_{h \to 0} h^{1-\beta} \frac{\partial}{\partial h} M_a(h)}_{N_h} + (1-\beta) \lim_{h \to 0} \frac{M_a(h)}{h^{\beta}}$$

Then we apply L'Hôpital's rule on the second term :

$$\Gamma(1-\beta) \mathcal{D}^{\beta}_{KG+} f(a) = \lim_{h \to 0} N_h + \frac{1-\beta}{\beta} \underbrace{\lim_{h \to 0} h^{1-\beta} \frac{\partial}{\partial h} M_a(h)}_{N_h} = \frac{1}{\beta} \lim_{h \to 0} N_h$$

Finally,

$$\mathcal{D}^{\beta}_{KG+} f(a) = \frac{1}{\beta \Gamma(1-\beta)} \lim_{h \to 0} h^{1-\beta} \frac{\partial}{\partial h} \int_0^1 \frac{f(hu+a) - f(a)}{(1-u)^{\beta}} du \tag{7}$$

From this equation there are two conclusions that can be drawn

First, for $f \in L^1(a, x)$ by application of the definition of fractional velocity and L'Hôpital's rule:

$$\mathcal{D}^{\beta}_{KG+} f(a) = \frac{v^{\beta}_{+} M_a(0)}{\Gamma(1-\beta)} \tag{8}$$

if the last limit exists. Therefore, LFD can be characterized in terms of fractional velocity. This can be formalized in the following proposition:

Proposition 7. *Suppose that $f \in \mathcal{I}^{\alpha}_{a,\pm}(L^1)$ for $x \in [a, a+\delta)$ (resp. $x \in (a-\delta, a]$) for some small $\delta > 0$. If $v^{\beta}_{\pm} M_a(0)$ exists finitely then*

$$\mathcal{D}^{\beta}_{KG\pm} f(a) = \frac{v^{\beta}_{\pm} M_a(0)}{\Gamma(1-\beta)}$$

where $M_a(h)$ is given by Formula (6).

From this we see that f may not be β-differentiable at x. From this perspective LFD is a derived concept - it is the $\beta-$ velocity of the integral average.

Second, for BV functions the order of integration and parametric derivation can be exchanged so that

$$\mathcal{D}^{\beta}_{KG+} f(a) = \frac{1}{\beta\Gamma(1-\beta)} \lim_{h\to 0} h^{1-\beta} \int_0^1 \frac{u f'(hu+a)}{(1-u)^{\beta}} du \tag{9}$$

where we demand the existence of $f'(x)$ a.e in (a, x), which follows from the Lebesgue differentiation theorem. This statement can be formalized as

Proposition 8. *Suppose that $f \in BV(a, \delta]$ for some small $\delta > 0$. Then*

$$\mathcal{D}^{\beta}_{KG+} f(a) = \frac{1}{\beta\Gamma(1-\beta)} \lim_{h\to 0} h^{1-\beta} \int_0^1 \frac{u f'(hu+a)}{(1-u)^{\beta}} du$$

In the last two formulas we can also set $\Gamma(-\beta) = \beta\Gamma(1-\beta)$ by the reflection formula. Therefore, in the conventional form for a BV function

$$\mathcal{D}^{\beta}_{KG+} f(a) = \frac{1}{\beta\Gamma(1-\beta)} \lim_{x\to a+} (x-a)^{1-\beta} \frac{\partial}{\partial x} \int_0^1 \frac{f((x-a)u+a) - f(a)}{(1-u)^{\beta}} du \tag{10}$$

5. Discussion

Kolwankar-Gangal local fractional derivative was introduced as tools for the study of the scaling of physical systems and systems exhibiting fractal behavior [39]. The conditions for applicability of the K-G fractional derivative were not specified in the seminal paper, which leaves space for different interpretations and sometimes confusions. For example, recently Tarasov claimed that local fractional derivatives of fractional order vanish everywhere [40]. In contrast, the results presented here demonstrate that local fractional derivatives vanish only if they are continuous. Moreover, they are non-zero on arbitrary dense sets of measure zero for β-differentiable functions as shown.

Another confusion is the initial claim presented in [23] that K-G fractional derivative is equivalent to what is called here β-fractional velocity. This needed to be clarified in [26] and restricted to the more limited functional space of summable fractional Riemann-Liouville derivatives [34].

Presented results call for a careful inspection of the claims branded under the name of "local fractional calculus" using K-G fractional derivative. Specifically, in the implied conditions on image function's regularity and arguments of continuity of resulting *local fractional derivative* must be examined in all cases. For example, in another stream of literature fractional difference quotients are defined on fractal sets, such as the Cantor's set [41]. This is not to be confused with the original approach of Cherebit, Kolwankar and Gangal where the topology is of the real line and the set χ_α is totally disconnected.

6. Conclusions

As demonstrated here, fractional velocities can be used to characterize the set of change of F-analytic functions. Local fractional derivatives and the equivalent fractional velocities have several distinct properties compared to integer-order derivatives. This may induce some wrong expectations to uninitiated reader. Some authors can even argue that these concepts are not suitable tools to deal with non-differentiable functions. However, this view pertains only to expectations transfered from the behavior of ordinary derivatives. On the contrary, one-sided local fractional derivatives can be used as a tool to study **local non-linear behavior** of functions as demonstrated by the presented examples. In applied problems, local fractional derivatives can be also used to derive fractional Taylor expansions [24,42,43].

Acknowledgments: The work has been supported in part by a grant from Research Fund—Flanders (FWO), contract number VS.097.16N. Graphs are prepared with the computer algebra system Maxima.

Conflicts of Interest: The author declares no conflict of interest.

Appendix A. General Definitions and Notations

The term *function* denotes a mapping $f : \mathbb{R} \mapsto \mathbb{R}$ (or in some cases $\mathbb{C} \mapsto \mathbb{C}$). The notation $f(x)$ is used to refer to the value of the mapping at the point x. The term *operator* denotes the mapping from functional expressions to functional expressions. Square brackets are used for the arguments of operators, while round brackets are used for the arguments of functions. $Dom[f]$ denotes the domain of definition of the function $f(x)$. The term Cauchy sequence will be always interpreted as a null sequence.

BVC[I] will mean that the function f is continuous of bounded variation (BV) in the interval I. AC[I] will mean that the function f is absolutely continuous in the interval I.

Definition A1. *A function $f(x)$ is called singular (SC) on the interval $x \in [a, b]$, if it is (i) non-constant; (ii) continuous; (iii) $f'(x) = 0$ Lebesgue almost everywhere (i.e., the set of non-differentiability of f is of measure 0) and (iv) $f(a) \neq f(b)$.*

There are inclusions $SC \subset BVC$ and $AC \subset BVC$.

Definition A2 (Asymptotic $\mathcal{O}$ notation)**.** *The notation $\mathcal{O}(x^\alpha)$ is interpreted as the convention that*

$$\lim_{x \to 0} \frac{\mathcal{O}(x^\alpha)}{x^\alpha} = 0$$

for $\alpha > 0$. The notation $\mathcal{O}_x$ will be interpreted to indicate a Cauchy-null sequence with no particular power dependence of x.

Definition A3. *We say that f is of (point-wise) Hölder class $\mathbb{H}^\beta$ if for a given x there exist two positive constants $C, \delta \in \mathbb{R}$ that for an arbitrary $y \in Dom[f]$ and given $|x - y| \leq \delta$ fulfill the inequality $|f(x) - f(y)| \leq C|x - y|^\beta$, where $|\cdot|$ denotes the norm of the argument.*

Further generalization of this concept will be given by introducing the concept of **F-analytic** functions.

Definition A4. *Consider a countable ordered set* $\mathbb{E}^{\pm} = \{\alpha_1 < \alpha_2 < \ldots\}$ *of positive real constants* α*. Then F-analytic* $\mathbb{F}^{E}$ *is a function which is defined by the convergent (fractional) power series*

$$F(x) := c_0 + \sum_{\alpha_i \in \mathbb{E}^{\pm}} c_i \left(\pm x + b_i\right)^{\alpha_i}$$

for some sets of constants $\{b_i\}$ *and* $\{c_i\}$*. The set* $\mathbb{E}^{\pm}$ *will be denoted further as the Hölder spectrum of* f *(i.e.,* $\mathbb{E}^{\pm}{}_f$*).*

Remark A1. *A similar definition is used in Oldham and Spanier [8], however, there the fractional exponents were considered to be only rational-valued for simplicity of the presented arguments. The minus sign in the formula corresponds to reflection about a particular point of interest. Without loss of generality only the plus sign convention is assumed.*

Definition A5. *Define the parametrized difference operators acting on a function* $f(x)$ *as*

$$\Delta^{+}_{\epsilon}\left[f\right](x) := f(x+\epsilon) - f(x)\,,$$
$$\Delta^{-}_{\epsilon}\left[f\right](x) := f(x) - f(x-\epsilon)$$

where $\epsilon > 0$*. The first one we refer to as forward difference operator, the second one we refer to as backward difference operator.*

The concept of point-wise oscillation is used to characterize the set of continuity of a function.

Definition A6. *Define forward oscillation and its limit as the operators*

$$\mathrm{osc}^{+}_{\epsilon}\left[f\right](x) := \sup_{[x,x+\epsilon]}\left[f\right] - \inf_{[x,x+\epsilon]}\left[f\right]$$
$$\mathrm{osc}^{+}[f](x) := \lim_{\epsilon\to 0}\left(\sup_{[x,x+\epsilon]} - \inf_{[x,x+\epsilon]}\right) f = \lim_{\epsilon\to 0}\mathrm{osc}^{+}_{\epsilon}\left[f\right](x)$$

and backward oscillation and its limit as the operators

$$\mathrm{osc}^{-}_{\epsilon}\left[f\right](x) := \sup_{[x-\epsilon,x]}\left[f\right] - \inf_{[x-\epsilon,x]}\left[f\right]$$
$$\mathrm{osc}^{-}[f](x) := \lim_{\epsilon\to 0}\left(\sup_{[x-\epsilon,x]} - \inf_{[x-\epsilon,x]}\right) f = \lim_{\epsilon\to 0}\mathrm{osc}^{-}_{\epsilon}\left[f\right](x)$$

according to previously introduced notation [44].

This definitions are used to identify two conditions, which help characterize fractional derivatives and velocities.

Appendix B. Essential Properties of Fractional Velocity

In this section we assume that the functions are BVC in the neighborhood of the point of interest. Under this assumption we have

- Product rule

$$v^{\beta}_{+}[f\,g](x) = v^{\beta}_{+}f(x)\,g(x) + v^{\beta}_{+}g(x)\,f(x) + [f,g]^{+}_{\beta}(x)$$
$$v^{\beta}_{-}[f\,g](x) = v^{\beta}_{-}f(x)\,g(x) + v^{\beta}_{-}g(x)\,f(x) - [f,g]^{-}_{\beta}(x)$$

- Quotient rule

$$v_{+}^{\beta}[f/g](x) = \frac{v_{+}^{\beta} f(x)\, g(x) - v_{+}^{\beta} g(x)\, f(x) - [f,g]_{\beta}^{+}}{g^2(x)}$$
$$v_{-}^{\beta}[f/g](x) = \frac{v_{-}^{\beta} f(x)\, g(x) - v_{-}^{\beta} g(x)\, f(x) + [f,g]_{\beta}^{-}}{g^2(x)}$$

where

$$[f,g]_{\beta}^{\pm}(x) := \lim_{\epsilon \to 0} v_{\epsilon\pm}^{\beta/2}[fg](x)$$

wherever the limit exists finitely.

For compositions of functions

- $f \in \mathbb{H}^{\beta}$ and $g \in \mathbb{C}^{1}$

$$v_{+}^{\beta} f \circ g(x) = v_{+}^{\beta} f(g)\, (g'(x))^{\beta}$$
$$v_{-}^{\beta} f \circ g(x) = v_{-}^{\beta} f(g)\, (g'(x))^{\beta}$$

- $f \in \mathbb{C}^{1}$ and $g \in \mathbb{H}^{\beta}$

$$v_{+}^{\beta} f \circ g(x) = f'(g)\, v_{+}^{\beta} g(x)$$
$$v_{-}^{\beta} f \circ g(x) = f'(g)\, v_{-}^{\beta} g(x)$$

Reflection formula

For $f(x) + f(a-x) = b$

$$v_{+}^{\beta} f(x) = v_{-}^{\beta} f(a-x)$$

Basic evaluation formula for absolutely continuous function [44]:

$$v_{\pm}^{\beta} f(x) = \frac{1}{\beta} \lim_{\epsilon \to 0} \epsilon^{1-\beta} f'(x \pm \epsilon)$$

References

1. Mandelbrot, B. *Fractal Geometry of Nature*; Henry Holt & Co.: New York, NY, USA, 1982.
2. Mandelbrot, B. *Les Objets Fractals: Forme, Hasard et Dimension*; Flammarion: Paris, France, 1989.
3. Metzler, R.; Klafter, J. The restaurant at the end of the random walk: Recent developments in the description of anomalous transport by fractional dynamics. *J. Phys. A Math. Gen.* **2004**, *37*, R161–R208.
4. Wheatcraft, S.W.; Meerschaert, M.M. Fractional conservation of mass. *Adv. Water Resour.* **2008**, *31*, 1377–1381.
5. Caputo, M.; Mainardi, F. Linear models of dissipation in anelastic solids. *Rivista del Nuovo Cimento* **1971**, *1*, 161–198.
6. Mainardi, F. Fractional Calculus: Some Basic Problems in Continuum and Statistical Mechanics. In *Fractals and Fractional Calculus in Continuum Mechanics*; Springer: Wien, Austria; New York, NY, USA, 1997; pp. 291–348.
7. Gorenflo, R.; Mainardi, F. Continuous time random walk, Mittag-Leffler waiting time and fractional diffusion: Mathematical aspects. In *Anomalous Transport*; Wiley-VCH Verlag GmbH & Co. KGaA: Weinheim, Germany, 2008; pp. 93–127.
8. Oldham, K.B.; Spanier, J.S. *The Fractional Calculus: Theory and Applications of Differentiation and Integration to Arbitrary Order*; Academic Press: New York, NY, USA, 1974.
9. Schroeder, M. *Fractals, Chaos, Power Laws: Minutes from an Infinite Paradise*; Dover Publications: Mineola, NY, USA, 1991.

10. Losa, G.; Nonnenmacher, T. Self-similarity and fractal irregularity in pathologic tissues. *Mod. Pathol.* **1996**, *9*, 174–182.
11. Darst, R.; Palagallo, J.; Price, T. *Curious Curves*; World Scientific Publishing Company: Singapore, 2009.
12. John Hutchinson. Fractals and self similarity. *Indiana Univ. Math. J.* **1981**, *30*, 713–747.
13. Mandelbro, B.B. *Intermittent Turbulence in Self-Similar Cascades: Divergence of High Moments and Dimension of the Carrier*; Springer: New York, NY, USA, 1999; pp. 317–357.
14. Meneveau, C.; Sreenivasan, K.R. Simple multifractal cascade model for fully developed turbulence. *Phys. Rev. Lett.* **1987**, *59*, 1424–1427.
15. Sreenivasan, K.R.; Meneveau, C. The fractal facets of turbulence. *J. Fluid Mech.* **1986**, *173*, 357–386.
16. Puente, C.; López, M.; Pinzón, J.; Angulo, J. The gaussian distribution revisited. *Adv. Appl. Probab.* **1996**, *28*, 500–524.
17. Nottale, L. Scale relativity and fractal space-time: Theory and applications. *Found. Sci.* **2010**, *15*, 101–152.
18. Cresson, J.; Pierret, F. Multiscale functions, scale dynamics, and applications to partial differential equations. *J. Math. Phys.* **2016**, *57*, 053504.
19. Cherbit, G. Local dimension, momentum and trajectories. In *Fractals, Non-Integral Dimensions and Applications*; John Wiley & Sons: Paris, France, 1991; pp. 231–238.
20. Prodanov, D. Characterization of strongly non-linear and singular functions by scale space analysis. *Chaos Solitons Fractals* **2016**, *93*, 14–19.
21. Du Bois-Reymond, P. Versuch einer classification der willkürlichen functionen reeller argumente nach ihren aenderungen in den kleinsten intervallen. *J. Reine Angew. Math.* **1875**, *79*, 21–37.
22. Faber, G. Über stetige funktionen. *Math. Ann.* **1909**, *66*, 81–94.
23. Ben Adda, F.; Cresson, J. About non-differentiable functions. *J. Math. Anal. Appl.* **2001**, *263*, 721–737.
24. Prodanov, D. Conditions for continuity of fractional velocity and existence of fractional Taylor expansions. *Chaos Solitons Fractals* **2017**, *102*, 236–244.
25. Odibat, Z.M.; Shawagfeh, N.T. Generalized Taylor's formula. *Appl. Math. Comput.* **2007**, *186*, 286–293.
26. Chen, Y.; Yan, Y.; Zhang, K. On the local fractional derivative. *J. Math. Anal. Appl.* **2010**, *362*, 17–33.
27. Lomnicki, Z.; Ulam, S. Sur la théorie de la mesure dans les espaces combinatoires et son application au calcul des probabilités i. variables indépendantes. *Fundam. Math.* 1934, *23*, 237–278.
28. Cesàro, E. Fonctions continues sans dérivée. *Arch. Math. Phys.* **1906**, *10*, 57–63.
29. Salem, R. On some singular monotonic functions which are strictly increasing. *Trans. Am. Math. Soc.* **1943**, *53*, 427–439.
30. Berg, L.; Krüppel, M. De rham's singular function and related functions. *Zeitschrift für Analysis und Ihre Anwendungen* **2000**, *19*, 227–237.
31. Neidinger, R. A fair-bold gambling function is simply singular. *Am. Math. Mon.* **2016**, *123*, 3–18.
32. Gillespie, D.T. The mathematics of Brownian motion and Johnson noise. *Am. J. Phys.* **1996**, *64*, 225–240.
33. Zili, M. On the mixed fractional brownian motion. *J. Appl. Math. Stoch. Anal.* **2006**, *2006*, 32435.
34. Ben Adda, F.; Cresson, J. Corrigendum to "About non-differentiable functions". *J. Math. Anal. Appl.* **2013**, *408*, 409–413.
35. Kolwankar, K.M.; Lévy Véhel, J. Measuring functions smoothness with local fractional derivatives. *Fract. Calc. Appl. Anal.* **2001**, *4*, 285–301.
36. Samko, S.; Kilbas, A.; Marichev, O. *Fractional Integrals and Derivatives: Theory and Applications*; Gordon and Breach: Yverdon, Switzerland, 1993.
37. De Rham, G. Sur quelques courbes definies par des equations fonctionnelles. *Rendiconti del Seminario Matematico Università e Politecnico di Torino* **1957**, *16*, 101–113.
38. Kolwankar, K.M.; Gangal, A.D. Fractional differentiability of nowhere differentiable functions and dimensions. *Chaos* **1996**, *6*, 505–513.
39. Kolwankar, K.M.; Gangal, A.D. Local fractional Fokker-Planck equation. *Phys. Rev. Lett.* **1998**, *80*, 214–217.
40. Tarasov, V.E. Local fractional derivatives of differentiable functions are integer-order derivatives or zero. *Int. J. Appl. Comput. Math.* **2016**, *2*, 195–201.
41. Yang, X.J.; Baleanu, D.; Srivastava, H.M. *Local Fractional Integral Transforms and Their Applications*; Academic Press: Cambridge, MA, USA, 2015.

42. Liu, Z.; Wang, T.; Gao, G. A local fractional Taylor expansion and its computation for insufficiently smooth functions. *East Asian J. Appl. Math.* **2015**, *5*, 176–191.
43. Prodanov, D. Regularization of derivatives on non-differentiable points. *J. Phys. Conf. Ser.* **2016**, *701*, 012031.
44. Prodanov, D. Fractional variation of Hölderian functions. *Fract. Calc. Appl. Anal.* **2015**, *18*, 580–602.

© 2018 by the authors. Licensee MDPI, Basel, Switzerland. This article is an open access article distributed under the terms and conditions of the Creative Commons Attribution (CC BY) license (http://creativecommons.org/licenses/by/4.0/).

fractal and fractional

MDPI

Article

Series Solution of the Pantograph Equation and Its Properties

Sachin Bhalekar [1,*] and Jayvant Patade [1,2]

1 Department of Mathematics, Shivaji University, Kolhapur 416004, India; jayvantpatade1195@gmail.com
2 Ashokrao Mane Group of Institution, Vathar, Kolhapur 416112, India
* Correspondence: sbb_maths@unishivaji.ac.in; Tel.: +91-231-260-9218

Received: 26 October 2017; Accepted: 30 November 2017; Published: 8 December 2017

Abstract: In this paper, we discuss the classical pantograph equation and its generalizations to include fractional order and the higher order case. The special functions are obtained from the series solution of these equations. We study different properties of these special functions and establish the relation with other functions. Further, we discuss some contiguous relations for these special functions.

Keywords: pantograph equation; proportional delay; fractional derivative; Gaussian binomial coefficient

MSC: 33E99, 34K06, 34K07

1. Introduction

Ordinary differential equations (ODE) are widely used by researchers to model various natural systems. However, it is observed that such equations cannot model the actual behavior of the system. Since the ordinary derivative is a local operator, it cannot model the memory and hereditary properties in real-life phenomena. Such phenomena can be modeled in a more accurate way by introducing some nonlocal component, e.g., delay in it.

The characteristic equation of delay differential equations (DDE) is a transcendental equation in contrast with the polynomial in the case of ODE. Hence, the DDEs are difficult to analyze as compared with ODEs.

Various special functions viz. exponential, sine, cosine, hypergeometric, Mittag-Leffler and gamma are obtained from ODEs [1]. If the equations have variable coefficients, then we may get the Legendre polynomial, the Laguerre polynomial, Bessel functions, and so on [2]. However, there are vary few papers that are devoted to the special functions arising in DDEs [3]. This motivates us to work on the special functions emerging from the solution of DDE with proportional delay. We analyze different properties of such special functions and present the relationship with other functions.

2. Preliminaries

Definition 1 ([4]). *Gaussian binomial coefficients are defined by:*

$$\binom{n}{r}_q = \begin{cases} \frac{(1-q^n)(1-q^{n-1})\cdots(1-q^{n-r+1})}{(1-q)(1-q^2)\cdots(1-q^r)} & \text{if } r \le n \\ 0 & \text{if } r > n, \end{cases} \tag{1}$$

where with $\mid q \mid < 1$.

Definition 2 ([5]). *The Riemann–Liouville integral of order* $\mu, \mu > 0$ *is given by:*

$$I^\mu f(t) = \frac{1}{\Gamma(\mu)} \int_0^t (t-\tau)^{\mu-1} f(\tau) d\tau, \quad t > 0. \tag{2}$$

Definition 3 ([5]). *The Caputo fractional derivative of f is defined as:*

$$\begin{aligned} D^{\mu} f(t) &= \frac{d^m}{dt^m} f(t), \quad \mu = m \\ &= I^{m-\mu} \frac{d^m}{dt^m} f(t), \quad m-1 < \mu < m, \quad m \in \mathbb{N}. \end{aligned} \tag{3}$$

Definition 4 ([6]). *Two special functions are said to be contiguous if their parameters differ by integers. The relations made by contiguous functions are said to be contiguous function relations.*

Theorem 1 ([7]). *(Existence and uniqueness theorem)*
Let X be a Banach space and $J = [0, T]$. If $f : J \times X \times X \to X$ is continuous and there exists a positive constant $L > 0$, such that $\| f(t,u,x) - f(t,v,y) \| \leq L(\| u - v \| + \| x - y \|)$, $t \in J$, $u, v, x, y \in X$ and if $\frac{4T^{\alpha}L}{\Gamma(\alpha+1)} < 1$, then the fractional differential equation with proportional delay:

$$D^{\alpha} u(t) = f(u, u(t), u(qt)), \quad u(0) = u_0, \quad t \in [0, T]$$

where $0 < \alpha \leq 1$, $0 < q < 1$ has a unique solution.

3. Pantograph Equation

The pantograph is a current collection device, which is used in electric trains. The mathematical model of the pantograph is discussed by various researchers [8–11].

The differential equation:

$$y'(t) = ay(t) + by(qt), \quad y(0) = 1, \tag{4}$$

with proportional delay modeling these phenomena is discussed by Ockendon and Tayler in [12]. Equation (4) is called the pantograph equation. Kato and McLeod [13] showed that the problem (4) is well-posed if $q < 1$. Further, the authors discussed the asymptotic properties of this equation. The coefficients in the power series solution are obtained by using a recurrence relation.

In [14], Fox et al. showed that the solution of (4) is given by a power series:

$$y(t) = 1 + \sum_{n=1}^{\infty} \frac{t^n}{n!} \prod_{j=0}^{n-1} \left(a + bq^j \right). \tag{5}$$

Iserles [15] considered a generalized pantograph equation $y'(t) = Ay(t) + By(qt) + Cy'(qt), y(0) = y_0$, where $q \in (0,1)$. The condition for the well-posedness of the problem is given in terms of A, B and C. The solution of the problem is expressed in the form of the Dirichlet series. Further, the advanced pantograph equation $y^{(n)}(t) = \sum_{j=0}^{l} \sum_{k=0}^{m-1} a_{j,k} y^{(k)}(\alpha_j t), t \geq 0$, where $a_{j,k} \in \mathbb{C}$ and $\alpha_j > 1$ for all $j = 0, 1, \ldots, l$, is also analyzed by Derfel and Iserles [16]. In [17], Patade and Bhalekar discussed the pantograph equation with incommensurate delay. Various properties of the series solution obtained are discussed.

In this paper, we write the solution (5) in the form of a special function and study its properties. We discuss the generalization of (4) to fractional order and the higher order case, as well.

4. Special Function Generated from the Pantograph Equation

We write solution (5) in the form of following special function:

$$\mathcal{R}(a, b, q, t) = 1 + \sum_{n=1}^{\infty} \frac{t^n}{n!} \prod_{j=0}^{n-1} \left(a + bq^j \right). \tag{6}$$

The notation $\mathcal{R}$ is used in memory of Ramanujan [18].

Theorem 2 ([14]). *If $q \in (0,1)$, then the power series:*

$$\mathcal{R}(a,b,q,t) = 1 + \sum_{n=1}^{\infty} \frac{t^n}{n!} \prod_{j=0}^{n-1} \left(a + bq^j\right)$$

has an infinite radius of convergence.

Corollary 1. *The power series (6) is absolutely convergent for all t, and hence, it is uniformly convergent on any compact interval on $\mathbb{R}$.*

Theorem 3. *For $m \in \mathbb{N} \cup \{0\}$, we have:*

$$\frac{d}{dt}\mathcal{R}(a,b,q,q^m t) = aq^m \mathcal{R}(a,b,q,q^m t) + bq^m \mathcal{R}(a,b,q,q^{m+1}t).$$

Proof. Consider:

$$\begin{aligned}
\frac{d}{dt}\mathcal{R}(a,b,q,q^m t) &= \frac{d}{dt}\left(1 + \sum_{n=1}^{\infty} \frac{(q^m t)^n}{n!} \prod_{j=0}^{n-1}\left(a + bq^j\right)\right) \\
&= q^m \sum_{n=1}^{\infty} \frac{(q^m t)^{n-1}}{(n-1)!} \prod_{j=0}^{n-1}\left(a + bq^j\right) \\
&= q^m (a+b) + q^m \sum_{n=2}^{\infty} \frac{(q^m t)^{n-1}}{(n-1)!} \prod_{j=0}^{n-1}\left(a + bq^j\right) \\
&= q^m (a+b) + q^m \sum_{n=1}^{\infty} \frac{(q^m t)^{n}}{n!} \prod_{j=0}^{n}\left(a + bq^j\right) \\
&= q^m (a+b) + q^m \sum_{n=1}^{\infty} \frac{(q^m t)^{n}(a+bq^n)}{n!} \prod_{j=0}^{n-1}\left(a + bq^j\right) \\
&= aq^m \left(1 + \sum_{n=1}^{\infty} \frac{(q^m t)^n}{n!} \prod_{j=0}^{n-1}\left(a + bq^j\right)\right) \\
&\quad + bq^m \left(1 + \sum_{n=1}^{\infty} \frac{(q^{m+1} t)^n}{n!} \prod_{j=0}^{n-1}\left(a + bq^j\right)\right) \\
\frac{d}{dt}\mathcal{R}(a,b,q,q^m t) &= aq^m \mathcal{R}(a,b,q,q^m t) + bq^m \mathcal{R}(a,b,q,q^{m+1}t)
\end{aligned}$$

Hence the proof. □

Theorem 4. *For $m \in \mathbb{N}$, we have:*

$$\frac{d^m}{dt^m}\mathcal{R}(a,b,q,t) = \sum_{r=0}^{m} q^{\binom{r}{2}} \binom{m}{r}_q a^{m-r} b^r \mathcal{R}(a,b,q,q^r t).$$

Proof. We prove the result by using the induction hypothesis on m.

From Theorem 3, the result is true for $m = 1$.

Suppose,

$$\frac{d^{m-1}}{dt^{m-1}}\mathcal{R}(a,b,q,t) = \sum_{r=0}^{m-1} q^{\binom{r}{2}} \binom{m-1}{r}_q a^{m-(r+1)} b^r \mathcal{R}(a,b,q,q^r t).$$

Consider,

$$\begin{aligned}
\frac{d^m}{dt^m}\mathcal{R}(a,b,q,t) &= \frac{d}{dt}\left(\frac{d^{m-1}}{dt^{m-1}}\mathcal{R}(a,b,q,t)\right) \\
&= \frac{d}{dt}\left(\sum_{r=0}^{m-1} q^{\binom{r}{2}} \binom{m-1}{r}_q a^{m-(r+1)} b^r \mathcal{R}(a,b,q,q^r t)\right) \\
&= \sum_{r=0}^{m-1} q^{\binom{r}{2}} \binom{m-1}{r}_q a^{m-(r+1)} b^r \frac{d}{dt}\mathcal{R}(a,b,q,q^r t).
\end{aligned}$$

Using Theorem 3, we have:

$$\begin{aligned}
\frac{d^m}{dt^m}\mathcal{R}(a,b,q,t) &= \sum_{r=0}^{m-1} q^{\binom{r}{2}} \binom{m-1}{r}_q a^{m-(r+1)} b^r \\
&\quad \left(aq^r\mathcal{R}(a,b,q,q^r t) + bq^r\mathcal{R}(a,b,q,q^{r+1}t)\right) \\
&= \sum_{r=0}^{m-1} q^{\binom{r}{2}} q^r \binom{m-1}{r}_q a^{m-r} b^r \mathcal{R}(a,b,q,q^r t) \\
&\quad + \sum_{k=0}^{m-1} q^{\binom{k}{2}} q^k \binom{m-1}{k}_q a^{m-(k+1)} b^{k+1} \mathcal{R}(a,b,q,q^{k+1}t) \\
&= \sum_{r=0}^{m-1} q^{\binom{r}{2}} q^r \binom{m-1}{r}_q a^{m-r} b^r \mathcal{R}(a,b,q,q^r t) \\
&\quad + \sum_{k=1}^{m} q^{\binom{k}{2}} \binom{m-1}{k-1}_q a^{m-k} b^k \mathcal{R}(a,b,q,q^k t) \\
&= \sum_{r=0}^{m} q^{\binom{r}{2}} q^r \binom{m-1}{r}_q a^{m-r} b^r \mathcal{R}(a,b,q,q^r t) \\
&\quad - q^{\binom{m}{2}} q^m \binom{m-1}{m}_q b^m \mathcal{R}(a,b,q,q^m t) \\
&\quad + \sum_{k=0}^{m} q^{\binom{k}{2}} \binom{m-1}{k-1}_q a^{m-k} b^k \mathcal{R}(a,b,q,q^k t) \\
&\quad - \binom{m-1}{-1}_q a^m \mathcal{R}(a,b,q,t) \\
&= \sum_{r=0}^{m} q^{\binom{r}{2}} q^r \binom{m-1}{r}_q a^{m-r} b^r \mathcal{R}(a,b,q,q^r t) \\
&\quad + \sum_{k=0}^{m} q^{\binom{k}{2}} \binom{m-1}{k-1}_q a^{m-k} b^k q^r \mathcal{R}(a,b,q,q^k t) \\
&= \sum_{r=0}^{m} q^{\binom{r}{2}} \left(q^r \binom{m-1}{r}_q + \binom{m-1}{r-1}_q\right) a^{m-r} b^r \mathcal{R}(a,b,q,q^r t).
\end{aligned}$$

$$\frac{d^m}{dt^m}\mathcal{R}(a,b,q,t) = \sum_{r=0}^{m} q^{\binom{r}{2}} \binom{m}{r}_q a^{m-r}b^r\mathcal{R}(a,b,q,q^r t),$$

This completes the proof. □

Theorem 5. *For $q \in (0,1)$, we have:*

$$\mathcal{R}(a,b,q,t) = 1 + \sum_{n=1}^{\infty}\sum_{r=0}^{n} q^{\binom{r}{2}} \binom{n}{r}_q a^{n-r}b^r \frac{t^n}{n!}.$$

Proof. We have:

$$\mathcal{R}(a,b,q,t) = 1 + \sum_{n=1}^{\infty}\frac{t^n}{n!}\prod_{j=0}^{n-1}\left(a+bq^j\right).$$

Using the q- binomial theorem:

$$\prod_{j=0}^{n-1}\left(a+bq^j\right) = \sum_{r=0}^{n} q^{\binom{r}{2}} \binom{n}{r}_q a^{n-r}b^r, \quad (7)$$

defined in [19], we obtain:

$$\mathcal{R}(a,b,q,t) = 1 + \sum_{n=1}^{\infty}\frac{t^n}{n!}\sum_{r=0}^{n} q^{\binom{r}{2}} \binom{n}{r}_q a^{n-r}b^r$$

$$\mathcal{R}(a,b,q,t) = 1 + \sum_{n=1}^{\infty}\sum_{r=0}^{n} q^{\binom{r}{2}} \binom{n}{r}_q a^{n-r}b^r \frac{t^n}{n!}.$$

Hence the proof. □

Theorem 6. *If $q \in (0,1)$, then we can express the integral of $e^{-t}\mathcal{R}$ in the following series containing $q-$binomial coefficients:*

$$\int_t^{\infty} e^{-t}\mathcal{R}(a,b,q,t)dt = \Gamma(1,t)\left(1 + \sum_{n=1}^{\infty}\sum_{k=0}^{n}\sum_{r=0}^{n} q^{\binom{r}{2}} \binom{n}{r}_q a^{n-r}b^r \frac{t^k}{k!}\right).$$

Proof. Consider:

$$\begin{aligned}
\int_t^{\infty} e^{-t}\mathcal{R}(a,b,q,t)dt &= \int_t^{\infty} e^{-t} + \sum_{n=1}^{\infty}\prod_{j=0}^{n-1}\frac{(a+bq^j)}{n!}\int_t^{\infty} e^{-t}t^n dt \\
&= e^{-t} + \sum_{n=1}^{\infty}\sum_{k=0}^{n}\prod_{j=0}^{n-1}\frac{(a+bq^j)}{n!}\Gamma(n+1,t) \\
&= e^{-t} + \sum_{n=1}^{\infty}\prod_{j=0}^{n-1}\frac{(a+bq^j)}{n!}n!e^{-t}\sum_{k=0}^{n}\frac{t^k}{k!} \\
&= e^{-t}\left(1 + \sum_{n=1}^{\infty}\sum_{r=0}^{n} q^{\binom{r}{2}} \binom{n}{r}_q a^{n-r}b^r \sum_{k=0}^{n}\frac{t^k}{k!}\right) \\
\int_t^{\infty} e^{-t}\mathcal{R}(a,b,q,t)dt &= \Gamma(1,t)\left(1 + \sum_{n=1}^{\infty}\sum_{k=0}^{n}\sum_{r=0}^{n} q^{\binom{r}{2}} \binom{n}{r}_q a^{n-r}b^r \frac{t^k}{k!}\right).
\end{aligned}$$

□

Theorem 7. *If $q \in (0,1)$, then we have the following relation:*

$$\begin{aligned}\int_0^t e^{-t}\mathcal{R}(a,b,q,t)dt &= 1+\sum_{n=1}^{\infty}\sum_{r=0}^{n} q^{\binom{r}{2}}\binom{n}{r}_q a^{n-r}b^r \\ &\quad -\Gamma(1,t)\left(1+\sum_{n=1}^{\infty}\sum_{k=0}^{n}\sum_{r=0}^{n} q^{\binom{r}{2}}\binom{n}{r}_q a^{n-r}b^r\frac{t^k}{k!}\right).\end{aligned}$$

Proof. Consider:

$$\begin{aligned}\int_0^t e^{-t}\mathcal{R}(a,b,q,t)dt &= \int_0^t e^{-t}dt+\sum_{n=1}^{\infty}\prod_{j=0}^{n-1}\frac{(a+bq^j)}{n!}\int_0^t e^{-t}t^n dt \\ &= 1-e^{-t}+\sum_{n=1}^{\infty}\prod_{j=0}^{n-1}\frac{(a+bq^j)}{n!}\gamma(n+1,t) \\ &= 1-e^{-t}+\sum_{n=1}^{\infty}\prod_{j=0}^{n-1}\frac{(a+bq^j)}{n!}n!\left(1-e^{-t}\sum_{k=0}^{n}\frac{t^k}{k!}\right) \\ &= 1+\sum_{n=1}^{\infty}\sum_{r=0}^{n} q^{\binom{r}{2}}\binom{n}{r}_q a^{n-r}b^r \\ &\quad -e^{-t}\left(1+\sum_{n=1}^{\infty}\sum_{k=0}^{n}\sum_{r=0}^{n} q^{\binom{r}{2}}\binom{n}{r}_q a^{n-r}b^r\frac{t^k}{k!}\right) \quad \text{(From (7))} \\ \int_0^t e^{-t}\mathcal{R}(a,b,q,t)dt &= 1+\sum_{n=1}^{\infty}\sum_{r=0}^{n} q^{\binom{r}{2}}\binom{n}{r}_q a^{n-r}b^r \\ &\quad -\Gamma(1,t)\left(1+\sum_{n=1}^{\infty}\sum_{k=0}^{n}\sum_{r=0}^{n} q^{\binom{r}{2}}\binom{n}{r}_q a^{n-r}b^r\frac{t^k}{k!}\right).\end{aligned}$$

□

Theorem 8. *The function $\mathcal{R}$ shows the following relationship with the lower incomplete gamma function γ:*

$$\int_0^t e^{-t}\mathcal{R}(a,b,q,\lambda t)dt = \gamma(1,t)+\lambda\sum_{n=1}^{\infty}\sum_{m=0}^{n+1}\prod_{j=0}^{n-1}\left(a+bq^j\right)\frac{\lambda^n}{n!}\frac{\gamma(n+m+1,t)}{m!}(1-\lambda)^m.$$

Proof. We have:

$$\begin{aligned}\int_0^t e^{-t}\mathcal{R}(a,b,q,\lambda t)dt &= 1-e^{-t}+\sum_{n=1}^{\infty}\prod_{j=0}^{n-1}\frac{(a+bq^j)}{n!}\int_0^t e^{-t}(\lambda t)^n dt \\ &= \gamma(1,t)+\sum_{n=1}^{\infty}\prod_{j=0}^{n-1}\frac{(a+bq^j)}{n!}\gamma(n+1,\lambda t).\end{aligned}$$

By using the property [20]: $\gamma(n,\lambda t)=\lambda^n\sum_{m=0}^{n}\frac{\gamma(n+m,t)}{m!}(1-\lambda)^m$, we get

$$\int_0^t e^{-t}\mathcal{R}(a,b,q,\lambda t)dt = \gamma(1,t)+\lambda\sum_{n=1}^{\infty}\sum_{m=0}^{n+1}\prod_{j=0}^{n-1}\left(a+bq^j\right)\frac{\lambda^n}{n!}\frac{\gamma(n+m+1,t)}{m!}(1-\lambda)^m.$$

□

Theorem 9. *The following integral gives the relation between the function $\mathcal{R}$ and the upper incomplete gamma function Γ:*

$$\int_t^{\infty} e^{-t}\mathcal{R}(a,b,q,\lambda t)dt = \Gamma(1,t) + \lambda \sum_{n=1}^{\infty}\sum_{m=0}^{n+1}\prod_{j=0}^{n-1}\left(a+bq^j\right)\frac{\lambda^n}{n!}\frac{\Gamma(n+m+1,t)}{m!}(1-\lambda)^m.$$

Proof.

$$\begin{aligned}\int_t^{\infty} e^{-t}\mathcal{R}(a,b,q,\lambda t)dt &= e^{-t} + \sum_{n=1}^{\infty}\prod_{j=0}^{n-1}\frac{(a+bq^j)}{n!}\int_t^{\infty} e^{-t}(\lambda t)^n dt \\ &= \Gamma(1,t) + \sum_{n=1}^{\infty}\prod_{j=0}^{n-1}\frac{(a+bq^j)}{n!}\Gamma(n+1,\lambda t).\end{aligned}$$

By using the property [20]: $\Gamma(n,\lambda t) = \lambda^n \sum_{m=0}^{n} \frac{\Gamma(n+m,t)}{m!}(1-\lambda)^m$, we obtain

$$\int_t^{\infty} e^{-t}\mathcal{R}(a,b,q,\lambda t)dt = \Gamma(1,t) + \lambda \sum_{n=1}^{\infty}\sum_{m=0}^{n+1}\prod_{j=0}^{n-1}\left(a+bq^j\right)\frac{\lambda^n}{n!}\frac{\Gamma(n+m+1,t)}{m!}(1-\lambda)^m.$$

□

5. Properties, Relations and Identities of $\mathcal{R}(a,b,q,t)$

We state the following properties of $\mathcal{R}(a,b,q,t)$ without proof.

5.1. Properties of $\mathcal{R}(a,b,q,t)$

For $l,m \in \mathbb{N}$:

1. $\mathcal{R}(0,\pm b^m,q,t) = \mathcal{R}(0,1,q,\pm b^m t)$.
   (a) $\mathcal{R}\left(0,\pm a^l b^{-m},q,t\right) = \mathcal{R}\left(0,1,q,\pm a^l b^{-m} t\right)$.
   (b) $\mathcal{R}(0,\pm b^{-m},q,t) = \mathcal{R}(0,1,q,\pm b^{-m} t)$.
2. $\mathcal{R}(a^m,a^m,q,t) = \mathcal{R}(1,1,q,a^m t)$.
   (a) $\mathcal{R}(\pm a^{-m},\pm a^m,q,t) = \mathcal{R}(1,a^{2m},q,\pm a^{-m} t)$.
   (b) $\mathcal{R}(\pm a^{m},\pm a^{-m},q,t) = \mathcal{R}(1,a^{-2m},q,\pm a^{m} t)$.
   (c) $\mathcal{R}(-a^{m},a^{-m},q,t) = \mathcal{R}(1,-a^{-2m},q,-a^{m} t)$.
   (d) $\mathcal{R}(a^{m},-a^{-m},q,t) = \mathcal{R}(1,-a^{-2m},q,a^{m} t)$.
   (e) $\mathcal{R}(-a^{-m},a^{m},q,t) = \mathcal{R}(1,-a^{2m},q,-a^{-m} t)$.
   (f) $\mathcal{R}(a^{-m},-a^{m},q,t) = \mathcal{R}(1,-a^{2m},q,a^{-m} t)$.
3. $\mathcal{R}\left(a^l,b^m,q,t\right) = \mathcal{R}\left(1,a^{-l}b^m,q,a^l t\right)$.
   (a) $\mathcal{R}\left(a^l b^{-m},b^m,q,t\right) = \mathcal{R}\left(1,a^{-l}b^{2m},q,a^l b^{-m} t\right)$.
   (b) $\mathcal{R}\left(a^{-l} b^{m},b^m,q,t\right) = \mathcal{R}\left(1,a^{l},q,a^{-l} b^{m} t\right)$.
   (c) $\mathcal{R}\left(a^l b^{-m},a^m,q,t\right) = \mathcal{R}\left(1,a^{m-l}b^{m},q,a^l b^{-m} t\right)$.
   (d) $\mathcal{R}\left(a^{-l} b^{m},a^m,q,t\right) = \mathcal{R}\left(1,a^{l+m}b^{-m},q,a^{-l} b^{m} t\right)$.

5.2. Relation to Other Functions

1. $\mathcal{R}(a,b,q,t) = 1 + \sum_{n=1}^{\infty} \frac{B(t,n,1)}{(n-1)!} \prod_{j=0}^{n-1} (a + bq^j)$.
2. $\mathcal{R}(a,b,q,t) = 1 + \sum_{n=1}^{\infty} \frac{B(t,n,1)}{B(n,1)n!} \prod_{j=0}^{n-1} (a + bq^j)$.
3. $\mathcal{R}(a,b,q,t) = 1 + \sum_{n=1}^{\infty} \frac{I_t(n,1)}{n!} \prod_{j=0}^{n-1} (a + bq^j)$,

where $B(t,n,1)$ and $I_t(n,1)$ are the incomplete beta function and regularized incomplete beta function [4].

5.3. Contiguous Relations of $\mathcal{R}(a,b,q,t)$

1. $\mathcal{R}(a \pm 1, b, q, t) = \mathcal{R}(1, (a \pm 1)^{-1} b, q, (a \pm 1)t)$.
2. $\mathcal{R}(a, b \pm 1, q, t) = \mathcal{R}(1, a^{-1}(b \pm 1), q, at)$.
3. $\mathcal{R}(a \pm 1,, b \pm 1, q, t) = \mathcal{R}(1, (a \pm 1)^{-1}(b \pm 1), q, (a \pm 1)t)$.

6. Generalizations

6.1. Generalizations to Include the Fractional Order Derivative

Consider the fractional delay differential equation with proportional delay:

$$D^{\alpha} y(t) = ay(t) + by(qt), \quad y(0) = 1, \tag{8}$$

where $0 < \alpha \leq 1$, $q \in (0,1)$, $a \in \mathbb{R}$ and $b \in \mathbb{R}$.

The solution of (8) is:

$$y(t) = 1 + \sum_{n=1}^{\infty} \frac{t^{\alpha n}}{\Gamma(\alpha n + 1)} \prod_{j=0}^{n-1} \left(a + bq^{\alpha j}\right). \tag{9}$$

We denote the series in (9) by:

$$\mathcal{R}_{\alpha}(a,b,q,t) = 1 + \sum_{n=1}^{\infty} \frac{t^{\alpha n}}{\Gamma(\alpha n + 1)} \prod_{j=0}^{n-1} \left(a + bq^{\alpha j}\right).$$

Theorem 10. *If $q \in (0,1)$, then the power series:*

$$\mathcal{R}_{\alpha}(a,b,q,t) = 1 + \sum_{n=1}^{\infty} \frac{t^{\alpha n}}{\Gamma(\alpha n + 1)} \prod_{j=0}^{n-1} \left(a + bq^{\alpha j}\right),$$

is convergent for all finite values of t.

Theorem 11. *For $q \in (0,1)$, $a \geq 0$ and $b \geq 0$, the function $\mathcal{R}_{\alpha}(a,b,q,t)$ satisfies the following inequality:*

$$E_{\alpha}(at^{\alpha}) \leq \mathcal{R}_{\alpha}(a,b,q,t) \leq E_{\alpha}((a+b)t^{\alpha}), \quad 0 \leq t < \infty.$$

Proof. Since $q \in (0,1)$, $a \geq 0$ and $b \geq 0$, we have:

$$\begin{aligned} \prod_{j=0}^{n-1} \left(a + bq^{\alpha j}\right) &\leq (a+b)^n \\ \Rightarrow \frac{t^{\alpha n}}{\Gamma(\alpha n + 1)} \prod_{j=0}^{n-1} \left(a + bq^{\alpha j}\right) &\leq \frac{t^{\alpha n}(a+b)^n}{\Gamma(\alpha n + 1)}. \end{aligned}$$

Taking summation over n, we get:

$$\mathcal{R}_\alpha(a,b,q,t) \le E_\alpha((a+b)t^\alpha), \quad 0 \le t < \infty. \tag{10}$$

Similarly, we have:

$$\begin{aligned} a^n &\le \prod_{j=0}^{n-1}\left(a+bq^{\alpha j}\right) \\ \Rightarrow E_\alpha(at^\alpha) &\le \mathcal{R}_\alpha(a,b,q,t), \quad 0 \le t < \infty. \end{aligned} \tag{11}$$

From (10) and (11), we get:

$$E_\alpha(at^\alpha) \le \mathcal{R}_\alpha(a,b,q,t) \le E_\alpha((a+b)t^\alpha), \quad 0 \le t < \infty.$$

□

The result is illustrated in Figure 1a,b.

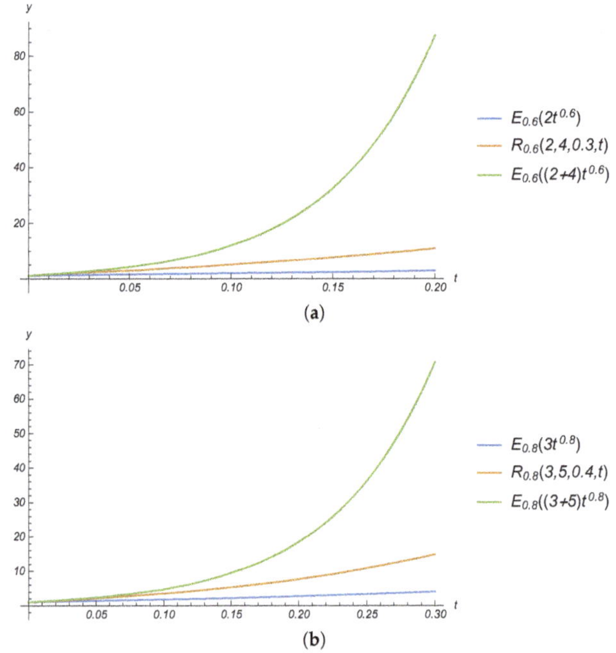

Figure 1. Bounds on $\mathcal{R}_\alpha$: (**a**) $\alpha = 0.6, a = 2, b = 4$ and $q = 0.3$; (**b**) $\alpha = 0.8, a = 3, b = 5$ and $q = 0.4$.

Theorem 12. *For $k \in \mathbb{N} \cup \{0\}$, we have:*

$$\frac{d^k}{dt^k}\mathcal{R}_\alpha(a,b,q,t) = \sum_{n=1}^{\infty} \frac{t^{\alpha n-k}}{\Gamma(\alpha n-k+1)} \prod_{j=0}^{n-1}\left(a+bq^{\alpha j}\right).$$

Theorem 13. *(Addition theorem)*

$$\mathcal{R}_\alpha(a,b,q,t+x) = \sum_{k=0}^{\infty} \frac{t^k}{k!}\mathcal{R}_\alpha^{(k)}(a,b,q,x),$$

where $\mathcal{R}_\alpha^{(k)}(a,b,q,x) = \frac{d^k}{dt^k}\mathcal{R}_\alpha(a,b,q,t)$

Proof. We have:

$$
\begin{aligned}
\mathcal{R}_\alpha(a,b,q,t+x) &= 1+\sum_{n=1}^{\infty}\frac{(t+x)^{\alpha n}}{\Gamma(\alpha n+1)}\prod_{j=0}^{n-1}\left(a+bq^{\alpha j}\right)\\
&= 1+\sum_{n=1}^{\infty}\frac{1}{\Gamma(\alpha n+1)}\left(x^{\alpha n}+\sum_{k=1}^{\infty}\frac{\Gamma(\alpha n+1)t^k x^{\alpha n-k}}{k!\Gamma(\alpha n-k+1)}\right)\prod_{j=0}^{n-1}\left(a+bq^{\alpha j}\right)\\
&= 1+\sum_{n=1}^{\infty}\frac{x^{\alpha n}}{\Gamma(\alpha n+1)}\prod_{j=0}^{n-1}\left(a+bq^{\alpha j}\right)\\
&\quad+\sum_{n=1}^{\infty}\sum_{k=1}^{\infty}\frac{t^k}{k!}\frac{x^{\alpha n-k}}{\Gamma(\alpha n-k+1)}\prod_{j=0}^{n-1}\left(a+bq^{\alpha j}\right)\\
&= \mathcal{R}_\alpha(a,b,q,x)+\sum_{k=1}^{\infty}\sum_{n=1}^{\infty}\frac{t^k}{k!}\frac{x^{\alpha n-k}}{\Gamma(\alpha n-k+1)}\prod_{j=0}^{n-1}\left(a+bq^{\alpha j}\right)\\
&= \mathcal{R}_\alpha(a,b,q,x)+\sum_{k=1}^{\infty}\frac{t^k}{k!}\sum_{n=1}^{\infty}\frac{x^{\alpha n-k}}{\Gamma(\alpha n-k+1)}\prod_{j=0}^{n-1}\left(a+bq^{\alpha j}\right)\\
&= \mathcal{R}_\alpha(a,b,q,x)+\sum_{k=1}^{\infty}\frac{t^k}{k!}\mathcal{R}_\alpha^{(k)}(a,b,q,x)\\
\mathcal{R}_\alpha(a,b,q,t+x) &= \sum_{k=0}^{\infty}\frac{t^k}{k!}\mathcal{R}_\alpha^{(k)}(a,b,q,x).
\end{aligned}
$$

□

Properties and Relations of $\mathcal{R}_\alpha(a,b,q,t)$

We get the following analogous properties and relations as in Section 5, e.g.,

1. $\mathcal{R}_\alpha\left(0,\pm b^{\alpha m},q,t\right) = \mathcal{R}_\alpha\left(0,1,q,\pm b^{\alpha m}t\right)$.
2. $\mathcal{R}_\alpha\left(a^{\alpha m},a^{\alpha m},q,t\right) = \mathcal{R}_\alpha\left(1,1,q,a^{\alpha m}t\right)$.
3. $\mathcal{R}_\alpha\left(a^{\alpha l},b^{\alpha m},q,t\right) = \mathcal{R}_\alpha\left(1,a^{-\alpha m}b^{\alpha m},q,a^{\alpha l}t\right)$ and so on.

6.2. Generalizations to the Higher Order Case

Consider the system of delay differential equation with proportional delay:

$$Y'(t) = AY(t)+BY(qt), \quad Y(0) = Y_0, \tag{12}$$

where $q \in (0,1)$, $A = \left(a_{ij}\right)_{n\times n}$, $B = \left(b_{ij}\right)_{n\times n}$ and $Y = \begin{bmatrix} y_1\\ y_2\\ \vdots\\ y_n\end{bmatrix}$.

The solution of (12) is:

$$\overline{\mathcal{R}}(A,B,q,tI) = \left(I+\sum_{n=1}^{\infty}\frac{t^n}{n!}\prod_{j=1}^{n}\left(A+Bq^{n-j}\right)\right)Y_0.$$

Theorem 14. *If $(A + Bq^n)$ is invertible for each n, then the power series:*

$$\overline{\mathcal{R}}(A, B, q, tI) = \left(I + \sum_{n=1}^{\infty} \frac{t^n}{n!} \prod_{j=1}^{n} \left(A + Bq^{n-j} \right) \right) Y_0,$$

where $A \in M_{n\times n}$ and $B \in M_{n\times n}$ is convergent for $t \in \mathbb{R}$.

6.3. Properties and Relations of $\overline{\mathcal{R}}(A, B, q, tI)$

We get the following analogous properties and relations as in Section 5 under the condition that A, B, $A \pm I$ and $B \pm I$ are invertible.

6.3.1. Properties of $\overline{\mathcal{R}}(A, B, q, tI)$

1. $\overline{\mathcal{R}}\left(0, \pm B^m, q, tI\right) = \overline{\mathcal{R}}\left(0, I, q, \pm B^m t\right)$.
2. $\overline{\mathcal{R}}\left(A^m, A^m, q, tI\right) = \overline{\mathcal{R}}\left(I, I, q, A^m t\right)$, etc.

6.3.2. Contiguous Relations of $\overline{\mathcal{R}}(A, B, q, tI)$

1. $\overline{\mathcal{R}}\left(A \pm I, B, q, tI\right) = \overline{\mathcal{R}}\left(I, (A \pm I)^{-1}B, q, (A \pm I)t\right)$.
2. $\overline{\mathcal{R}}\left(A, B \pm I, q, tI\right) = \overline{\mathcal{R}}\left(I, A^{-1}(B \pm I), q, At\right)$.
3. $\overline{\mathcal{R}}\left(A \pm I, , B \pm I, q, tI\right) = \overline{\mathcal{R}}\left(I, (A \pm I)^{-1}(B \pm I), q, (A \pm I)t\right)$.

7. Conclusions

In this article, we discussed the pantograph equation and its generalizations. The series solutions of these equations are treated as special functions. These special functions are different from hyper-geometric and other special functions because they are obtained as a solution of the delay differential equation. It is observed that the m-th derivative of $\mathcal{R}$ can be represented as a linear combination of $\mathcal{R}(a, b, q, q^r t)$ for $0 \leq r \leq m$ with Gaussian binomial coefficients. The function $\mathcal{R}_\alpha$ is shown to be bounded by the Mittag-Leffler functions. We have studied different properties and discussed some relations of these special functions. We hope that our work will encourage researchers to dig for more properties of the special functions obtained from delay differential equations.

Acknowledgments: Sachin Bhalekar acknowledges Council of Scientific and Industrial Research (CSIR), New Delhi, for funding through Research Project No. 25(0245)/15/EMR-II.

Author Contributions: Sachin Bhalekar suggested the problem, guided for the research work and confirmed the results. Jayvant Patade obtained the solution to the problem and derived the relations.

Conflicts of Interest: The authors declare no conflict of interest.

References

1. Mathai, A.M.; Haubold, H.J. *Special Functions for Applied Scientists*; Springer: New York, NY, USA, 2008.
2. Bell, W.W. *Special Functions for Scientists and Engineers*; Courier Corporation: London, UK, 2004.
3. Corless, R.M.; Gonnet, G.H.; Hare, D.E.G.; Jeffrey, D.J.; Knuth, D.E. On the Lambert W function. *Adv. Comput. Math.* **1996**, *5*, 329–359.
4. Magnus, W.; Oberhettinger, F.; Soni, R.P. *Formulas and Theorems for the Special Functions of Mathematical Physics*; Springer: New York, NY, USA, 2013; Volume 52.
5. Kilbas, A.A.; Srivastava, H.M.; Trujillo, J.J. *Theory and Applications of Fractional Differential Equations*; Elsevier: Amsterdam, The Netherlands, 2006.
6. Erdelyi, A.; Magnus, W.; Oberhettinger, F.; Tricomi, F.G. *Higher Transcendental Functions*; McGraw–Hill: New York, NY, USA, 1955.
7. Balachandran, K.; Kiruthika, S.; Trujillo, J.J. Existence of solutions of nonlinear fractional pantograph equations. *Acta Math. Sci.* **2013**, *33*, 712–720.

8. Andrews, H.I. Third paper: Calculating the behaviour of an overhead catenary system for railway electrification. *Proc. Inst. Mech. Eng.* **1964**, *179*, 809–846.
9. Abbott, M.R. Numerical method for calculating the dynamic behaviour of a trolley wire overhead contact system for electric railways. *Comput. J.* **1970**, *13*, 363–368.
10. Gilbert, G.; Davtcs, H.E.H. Pantograph motion on a nearly uniform railway overhead line. *Proc. Inst. Electr. Eng.* **1966**, *113*, 485–492.
11. Caine, P.M.; Scott, P.R. Single-wire railway overhead system. *Proc. Inst. Electr. Eng.* **1969**, *116*, 1217–1221.
12. Ockendon, J.; Tayler, A.B. The dynamics of a current collection system for an electric locomotive. *Proc. R. Soc. Lond. A Math. Phys. Eng. Sci.* **1971**, *322*, 447–468.
13. Kato, T.; McLeod, J.B. The functional-differential equation $y'(x) = \lambda y(qx) + by(x)$. *Bull. Am. Math. Soc.* **1971**, *77*, 891–937.
14. Fox, L.; Mayers, D.; Ockendon, J.R.; Tayler, A.B. On a functional differential equation. *IMA J. Appl. Math.* **1971**, *8*, 271–307.
15. Iserles, A. On the generalized pantograph functional-differential equation. *Eur. J. Appl. Math.* **1993**, *4*, 1–38.
16. Derfel, G.; Iserles, A. The pantograph equation in the complex plane. *J. Math. Anal. Appl.* **1997**, *213*, 117–132.
17. Patade, J.; Bhalekar, S. Analytical Solution of Pantograph Equation with Incommensurate Delay. *Phys. Sci. Rev.* **2017**, *2*, doi:10.1515/psr-2016-0103.
18. Kanigel, R. *The Man Who Knew Infinity*; Washington Square Press: New York, NY, USA, 2015.
19. Stanley, R.P. *Enumerative Combinatorics*; Cambridge University Press: Cambridge, UK, 1997.
20. Gautschi, W.; Harris, F.E.; Temme, N.M. Expansions of the exponential integral in incomplete gamma functions. *Appl. Math. Lett.* **2003**, *16*, 1095–1099.

© 2017 by the authors. Licensee MDPI, Basel, Switzerland. This article is an open access article distributed under the terms and conditions of the Creative Commons Attribution (CC BY) license (http://creativecommons.org/licenses/by/4.0/).

fractal and fractional

MDPI

Article

European Vanilla Option Pricing Model of Fractional Order without Singular Kernel

Mehmet Yavuz [1,*] and Necati Özdemir [2]

1 Department of Mathematics-Computer Sciences, Faculty of Science, Necmettin Erbakan University, Konya 42090, Turkey

2 Department of Mathematics, Faculty of Sciences and Arts, Balıkesir University, Balıkesir 10145, Turkey; nozdemir@balikesir.edu.tr

* Correspondence: mehmetyavuz@konya.edu.tr; Tel.: +90-332-323-8220 (ext. 5548)

Received: 28 December 2017; Accepted: 14 January 2018; Published: 16 January 2018

Abstract: Recently, fractional differential equations (FDEs) have attracted much more attention in modeling real-life problems. Since most FDEs do not have exact solutions, numerical solution methods are used commonly. Therefore, in this study, we have demonstrated a novel approximate-analytical solution method, which is called the Laplace homotopy analysis method (LHAM) using the Caputo–Fabrizio (CF) fractional derivative operator. The recommended method is obtained by combining Laplace transform (LT) and the homotopy analysis method (HAM). We have used the fractional operator suggested by Caputo and Fabrizio in 2015 based on the exponential kernel. We have considered the LHAM with this derivative in order to obtain the solutions of the fractional Black–Scholes equations (FBSEs) with the initial conditions. In addition to this, the convergence and stability analysis of the model have been constructed. According to the results of this study, it can be concluded that the LHAM in the sense of the CF fractional derivative is an effective and accurate method, which is computable in the series easily in a short time.

Keywords: fractional option pricing problem; Caputo–Fabrizio fractional derivative; homotopy analysis method; Laplace transform

---

1. Introduction and Some Preliminaries

Modeling with fractional calculus has become increasingly important in recent years. During the last few decades especially, new fractional operators, methods and algorithms have been developed relating to mathematical modeling and simulation. Many studies have been undertaken in the last quarter of a century on new fractional derivative operators. Fractional calculus (FC) and numerical-approximate solution methods are extensively used in the solution of real-life problems, such as mathematical, engineering, financial, biological and physical problems.

For example, these techniques have been used to evaluate the performance of an electrical resistance inductance and capacitance (RLC) circuit using a new fractional operator with a local and nonlocal kernel [1], to price fractional European vanilla-type options [2,3], to analyze a new model of H1N1 spread [4], to model the population growth [5], to apply the homotopy analysis method [6], the Adomian decomposition method [7,8], the homotopy perturbation method [9,10], He's variational iteration method in conformable derivative sense [11], the generalized differential transform method [12], the finite difference method [13] and the multivariate Padé approximation method [14]. Moreover, these proposed fractional techniques have been used to obtain the solution of the optimal control problem [15], the constrained optimization problem [16], the portfolio optimization problem [17], the diffusion-wave problem [18], etc.

In 2015, Caputo and Fabrizio developed a new fractional derivative operator built upon the exponential function to overcome the singular kernel problem [19]. Their fractional derivative has a smooth kernel, which takes on two different representations for the temporal and spatial variable. Atangana and Alkahtani [20] applied the CF derivative operator to the groundwater flowing within a confined aquifer. Singh et al. [21] analyzed the ENSO model in the global climate with the CF operator, and their simulations found out that when α tends to one, the CF derivative shows more interesting behavior. Ali et al. [22] obtained the solution of the fractional model of Walters'-B fluid by using CF fractional derivative. In another study, Morales-Delgado et al. [23] compared the solutions obtained by using the CF derivative and the Liouville–Caputo derivative. Sheikh et al. [24] used the analysis of the Atangana–Baleanu (AB) and CF for generalized Casson fluid model. Atangana and Alkahtani [20] modeled the groundwater flowing within a confined aquifer by using the CF derivative. Koca and Atangana [25] solved the Cattaneo–Hristov model with CF and AB operators. Also in [26–30], the authors studied some interesting problems based on the CF fractional derivative.

Many powerful approximate-analytical methods have been presented in the finance literature, especially in modeling the European and American option prices. For example, [9,31,32] are relatively new approaches providing an analytical and numerical approximation to the Black–Scholes option pricing equation. The financial system can be viewed as money, capital and derivative markets (options, futures, forwards, swaps, etc.). Options are widely used in global financial markets. An option is a right, but not an obligation. The most important benefit of the option is the ability to invest in large amounts with a very small capital.

In 1973, Fisher Black and Myron Scholes [33] investigated in their study a model that can easily compute the prices of the options. This model also can evaluate the Greeks of the options and ratio of hedge. The Black–Scholes model that prices stock options has been applied to many different possessions and payments. This form of the pricing model is one of the most meaningful mathematical equations for a financial instrument. The Black–Scholes model with respect to an option can be considered as [34,35]:

$$\frac{\partial V}{\partial t}+\frac{1}{2}\sigma^2 S^2\frac{\partial^2 V}{\partial S^2}+r\left(t\right)S\frac{\partial V}{\partial S}-r\left(t\right)V=0,\quad \left(S,t\right)\in \mathrm{R}^{+}\times\left(0,T\right), \tag{1}$$

where $V=V\left(S,t\right)$ shows the vanilla-type option price at asset price S and time t. T represents the maturity time; $r\left(t\right)$ is the risk-free interest rate; and $\sigma\left(S,t\right)$ is the volatility function of the underlying asset. In Equation (1), we observe that $V\left(0,t\right)=0$, $V\left(S,t\right)\sim S$ as $S\to\infty$, and we can write payoff functions as: $V_c\left(S,T\right)=\max\left(S-E,0\right)$ and $V_p\left(S,T\right)=\max\left(E-S,0\right)$, where $V_c\left(S,T\right)$ and $V_p\left(S,T\right)$ show the value of the vanilla call and put options, respectively, and E is the exercise (strike) price. The closed form solution of Equation (1) can be obtained by using the heat equation. In order to obtain the FBSE, we make the following conversions:

$$S=Ee^x,\ t=T-\frac{2\tau}{\sigma^2},\ V=E\omega\left(x,\tau\right).$$

This yields the equation:

$$\frac{\partial^\alpha\omega\left(x,\tau\right)}{\partial\tau^\alpha}=\frac{\partial^2\omega\left(x,\tau\right)}{\partial x^2}+\left(k-1\right)\frac{\partial\omega\left(x,\tau\right)}{\partial x}-k\omega\left(x,\tau\right),\ \tau>0,\ x\in R,\ 0<\alpha\leq 1, \tag{2}$$

with initial condition:

$$\omega\left(x,0\right)=\max\left(e^x-1,0\right). \tag{3}$$

Equation (2) is called the Black–Scholes option pricing equation of fractional order. In Equation (2), we define $k=2r/\sigma^2$, where k represents the balance between the interest rates' and stock returns' variability.

In addition to this, Cen and Le (2011) obtained the generalized fractional Black–Scholes equation [36] (GFBSE) by considering $r = 0.06$ and $\sigma = 0.4\,(2 + \sin x)$ in Equation (2):

$$\frac{\partial^\alpha \omega}{\partial \tau^\alpha} + 0.08\,(2 + \sin x)^2 \frac{\partial^2 \omega}{\partial x^2} + 0.06x\frac{\partial \omega}{\partial x} - 0.06\omega = 0,\ \tau > 0,\ x \in R,\ 0 < \alpha \leq 1, \tag{4}$$

with the initial condition:

$$\omega\,(x,0) = \max\left(x - 25e^{-0.06}, 0\right). \tag{5}$$

When the market mechanism is given a full scope, it is concluded that the estimating effect of the fractional Black–Scholes model would be more effective than the traditional Black–Scholes model. For example, according to a special study [37] on China Merchants Bank, it was found that Fractional Black–Scholes (FBS) price is bigger than the Black–Scholes (BS) price. Additionally, FBS is better than BS, while the volatility is relatively larger ($\sigma > 0.17$). We have aimed in this study to display the solution of fractional Black–Scholes Equations (2)–(5) using the proposed fractional derivative operator. Furthermore, we have aimed to determine the stability analysis of the method and the effectiveness of the CF operator using the results obtained.

Definition 1. *The usual Caputo time-fractional derivative of order α is given by [19]:*

$$D_t^\alpha f\,(t) = \frac{1}{\Gamma\,(1-\alpha)} \int_a^t \frac{f'\,(\lambda)}{(t-\lambda)^\alpha} d\lambda,\ 0 \leq \alpha \leq 1,\ a \in [-\infty, t)\,,\ f \in H^1\,(a,b)\,,\ b > a. \tag{6}$$

By changing the kernel $(t-\lambda)^{-\alpha}$ with the function $\exp\left(\frac{\alpha t}{\alpha-1}\right)$ and $\frac{1}{\Gamma(1-\alpha)}$ with $\frac{M(\alpha)}{\Gamma(1-\alpha)}$, we obtain the following new fractional time derivative named the Caputo–Fabrizio time fractional derivative.

Definition 2. *The definition of the CF sense derivative is given by [19]:*

$${}_0^{CF}D_t^\alpha f\,(t) = \frac{M\,(\alpha)}{1-\alpha} \int_a^t \exp\left[-\frac{\alpha\,(t-\lambda)}{1-\alpha}\right] d\lambda, \tag{7}$$

where $M\,(\alpha)$ is a normalization function such that $M\,(0) = M\,(1) = 1$. This definition can also be considered for functions that do not belong to $H^1\,(a,b)$, and the kernel has non-singularity for $t = \tau$. Equation (7) can be formulated also for $f \in L^1\,(-\infty, b)$ and for any $0 \leq \alpha \leq 1$ as:

$${}_0^{CF}D_t^\alpha f\,(t) = \frac{\alpha M\,(\alpha)}{1-\alpha} \int_{-\infty}^t (f\,(t) - f\,(\lambda)) \exp\left[-\frac{\alpha\,(t-\lambda)}{1-\alpha}\right] d\lambda. \tag{8}$$

Definition 3. *The Laplace transform of CF fractional derivative ${}_0^{CF}D_t^\alpha f\,(t)$ can be defined as follows [19]:*

$$\begin{aligned} L\left\{{}_0^{CF}D_t^{\alpha+n} f\,(t)\right\}(s) &= \tfrac{1}{1-\alpha} L\left\{f^{(\alpha+n)}\,(t)\right\} L\left\{\exp\left[-\tfrac{\alpha t}{1-\alpha}\right]\right\} \\ &= \tfrac{s^{n+1}L\{f(t)\} - s^n f(0) - s^{n-1}f'(0) - \cdots - f^{(n)}(0)}{s+\alpha(1-s)}. \end{aligned} \tag{9}$$

From Definition 3, we get the following special cases:

$$\begin{aligned} L\left\{{}_0^{CF}D_t^\alpha f\,(t)\right\}(s) &= \tfrac{sL\{f(t)\} - f(0)}{s+\alpha(1-s)},\ n = 0, \\ L\left\{{}_0^{CF}D_t^{\alpha+1} f\,(t)\right\}(s) &= \tfrac{s^2L\{f(t)\} - sf(0) - f'(0)}{s+\alpha(1-s)},\ n = 1. \end{aligned} \tag{10}$$

2. Description of the Method Using the Caputo–Fabrizio Fractional Operator

In this section of the study, we have demonstrated the solution method described by using the CF operator. Consider the following fractional PDE [5]:

$$ {}_{0}^{CF}D_t^{\alpha}\omega(x,t)+\eta(x)\frac{\partial\omega(x,t)}{\partial x}+\gamma(x)\frac{\partial^2\omega(x,t)}{\partial x^2}+\varphi(x)\omega(x,t)=\nu(x,t), \tag{11} $$

where $(x,t)\in[0,1]\times[0,T]$, with the initial conditions:

$$ \frac{\partial^k\omega}{\partial t^k}(x,0)=f_k(x),\ k=0,1,...,m-1, \tag{12} $$

and the boundary conditions:

$$ \omega(0,t)=g_0(t),\ \omega(1,t)=g_1(t),\ t\geq 0, \tag{13} $$

where $f_k, k=0,1,...,m-1, \nu, g_0, g_1, \eta, \gamma$ and φ are known functions and $T>0$ is a real number and $m-1<\alpha+n\leq m$. We define the method of solution for solving Problems (2)–(5). The LT of the CF derivative is satisfied as:

$$ L\left\{{}_{0}^{CF}D_t^{\alpha+n}\omega(x,t)\right\}=\frac{s^{n+1}L\{\omega(x,t)\}-s^n\omega(x,0)-s^{n-1}\omega'(x,0)-\cdots-\omega^{(n)}(x,0)}{s+\alpha(1-s)}. \tag{14} $$

In Equation (14), $s\geq 0$, and let us define the $L\{\omega(x,t)\}(s)=\Omega(x,s)$ for Equation (11), then we can write:

$$ \begin{aligned} \Omega(x,s)&=\left(\tfrac{\alpha(s-1)-s}{s^{n+1}}\right)\left[\eta(x)\tfrac{\partial}{\partial x}+\gamma(x)\tfrac{\partial^2}{\partial x^2}+\varphi(x)\right]\Omega(x,s)\\ &+\tfrac{1}{s^{n+1}}\left[s^n\omega_0(x)+s^{n-1}\omega_1(x)+\cdots+\omega_n(x)\right]+\tfrac{s+\alpha(1-s)}{s^{n+1}}\tilde{\nu}(x,s). \end{aligned} \tag{15} $$

Now, we can construct the homotopy for Equation (15) as follows:

$$ \begin{aligned} \Omega(x,s)&=z\left(\tfrac{\alpha(s-1)-s}{s^{n+1}}\right)\left[\eta(x)\tfrac{\partial}{\partial x}+\gamma(x)\tfrac{\partial^2}{\partial x^2}+\varphi(x)\right]\Omega(x,s)\\ &+\tfrac{1}{s^{n+1}}\left[s^n\omega_0(x)+s^{n-1}\omega_1(x)+\cdots+\omega_n(x)\right]+\tfrac{s+\alpha(1-s)}{s^{n+1}}\tilde{\nu}(x,s), \end{aligned} \tag{16} $$

where $\Omega(x,s)=L\{\omega(x,t)\}$ and $\tilde{\nu}(x,s)=L\{\nu(x,t)\}$. Furthermore, the Laplace transforms of the initial conditions are obtained as:

$$ \Omega(0,s)=L\{g_0(t)\}\ ,\Omega(1,s)=L\{g_1(t)\},\ s\geq 0. \tag{17} $$

Then, the solution of Equation (16) can be represented as:

$$ \Omega(x,s)=\sum_{m=0}^{\infty}z^m\Omega_m(x,s),\ m=0,1,2,\ldots. \tag{18} $$

Substituting Equation (18) into Equation (16), we have:

$$ \begin{aligned} \textstyle\sum_{m=0}^{\infty}z^m\Omega_m(x,s)&=z\left(\tfrac{\alpha(s-1)-s}{s^{n+1}}\right)\left[\eta(x)\tfrac{\partial}{\partial x}+\gamma(x)\tfrac{\partial^2}{\partial x^2}+\varphi(x)\right]\textstyle\sum_{m=0}^{\infty}z^m\Omega_m(x,s)\\ &+\tfrac{1}{s^{n+1}}\left[s^n\omega_0(x)+s^{n-1}\omega_1(x)+\cdots+\omega_n(x)\right]+\tfrac{s+\alpha(1-s)}{s^{n+1}}\tilde{\nu}(x,s). \end{aligned} \tag{19} $$

By comparing the coefficients of powers of z, we obtain the homotopies as follows:

$$ \begin{aligned} &z^0:\Omega_0(x,s)=\tfrac{1}{s^{n+1}}\left(s^n\omega_0(x)+s^{n-1}\omega_1(x)+\cdots+\omega_n(x)\right)+\left(\tfrac{s+\alpha(1-s)}{s^{n+1}}\right)\tilde{\nu}(x,s),\\ &z^1:\Omega_1(x,s)=-\left(\tfrac{s+\alpha(1-s)}{s^{n+1}}\right)\left[\eta(x)\tfrac{\partial}{\partial x}+\gamma(x)\tfrac{\partial^2}{\partial x^2}+\varphi(x)\right]\Omega_0(x,s),\\ &z^2:\Omega_2(x,s)=-\left(\tfrac{s+\alpha(1-s)}{s^{n+1}}\right)\left[\eta(x)\tfrac{\partial}{\partial x}+\gamma(x)\tfrac{\partial^2}{\partial x^2}+\varphi(x)\right]\Omega_1(x,s),\\ &\vdots\\ &z^{n+1}:\Omega_{n+1}(x,s)=-\left(\tfrac{s+\alpha(1-s)}{s^{n+1}}\right)\left[\eta(x)\tfrac{\partial}{\partial x}+\gamma(x)\tfrac{\partial^2}{\partial x^2}+\varphi(x)\right]\Omega_n(x,s). \end{aligned} \tag{20} $$

When the $z \to 1$, we see that Equation (20) gives the approximate solution for the problems (15) and (16), and the solution is given by:

$$T_n(x,s) = \sum_{j=0}^{n} \Omega_j(x,s). \tag{21}$$

If we take the inverse LT of Equation (21), we have the approximate solution of Equation (11),

$$\omega_{approx}(x,t) \cong \omega_n(x,t) = L^{-1}\{T_n(x,s)\}. \tag{22}$$

Furthermore, we will show the error rates of the solution with LHAM described above, by applying this method to the homogeneous fractional option pricing problem. If we define $\omega_n(x,\tau) = L^{-1}\{T_n(x,s)\}$, which is the first n-th sum of the series in the approximate solution of (22), the rate of absolute error R_{AE} is computed as:

$$R_{AE}(\%) = \left| \frac{\omega_n(x,t) - \omega_{exact}(x,t)}{\omega_{exact}(x,t)} \right| \times 100. \tag{23}$$

3. Solution of the European Option Pricing Problem

In this part of the study, we have solved the fractional Black–Scholes equation and generalized fractional Black–Scholes equation (FBSE), which are two of the most important option pricing models. We have regarded this as the method LHAM, which is described with the Caputo–Fabrizio fractional derivative.

3.1. Fractional European Option Pricing Problem in the Sense of the Caputo–Fabrizio Derivative

Now, we consider the classical FBSE (2) with the initial condition (3). Firstly, we solve this equation by using the LHAM in the sense of the CF fractional derivative operator. Because the equation is homogeneous, we obtain the LT of the right side of the equation as zero, i.e., $\tilde{v}(x,s) = L\{v(x,\tau)\} = 0$. Now, we create the homotopies as follows:

$$\begin{aligned}
z^0 : \Omega_0(x,s) &= \tfrac{1}{s} u(x,0) + \left(\tfrac{s+\alpha(1-s)}{s}\right)(0) = \tfrac{\max(e^x-1,0)}{s}, \\
z^1 : \Omega_1(x,s) &= \left(\tfrac{s+\alpha(1-s)}{s}\right)\left[\tfrac{\partial^2 \Omega_0(x,s)}{\partial x^2} + (k-1)\tfrac{\partial \Omega_0(x,s)}{\partial x} - k\Omega_0(x,s)\right] \\
&= \tfrac{k}{s}\left(\tfrac{s+\alpha(1-s)}{s}\right)(e^x - \max(e^x-1,0)), \\
z^2 : \Omega_2(x,s) &= \left(\tfrac{s+\alpha(1-s)}{s}\right)\left[\tfrac{\partial^2 \Omega_1(x,s)}{\partial x^2} + (k-1)\tfrac{\partial \Omega_1(x,s)}{\partial x} - k\Omega_1(x,s)\right] \\
&= -\tfrac{k^2}{s}\left(\tfrac{s+\alpha(1-s)}{s}\right)^2(e^x - \max(e^x-1,0)), \\
&\vdots \\
z^n : \Omega_n(x,s) &= \left(\tfrac{s+\alpha(1-s)}{s}\right)\left[\tfrac{\partial^2 \Omega_{n-1}(x,s)}{\partial x^2} + (k-1)\tfrac{\partial \Omega_{n-1}(x,s)}{\partial x} - k\Omega_{n-1}(x,s)\right] \\
&= (-1)^{n+1}\tfrac{k^n}{s}\left(\tfrac{s+\alpha(1-s)}{s}\right)^n(e^x - \max(e^x-1,0)).
\end{aligned} \tag{24}$$

By summing the iteration term up to n-th order, we obtain:

$$T_n(x,s) = \sum_{j=0}^{n} \Omega_j(x,s) = \frac{\max(e^x-1,0)}{s} + \frac{e^x - \max(e^x-1,0)}{s} \sum_{m=1}^{n} (k(s+\alpha(1-s)))^m. \tag{25}$$

Getting the inverse LT of Equation (25), we have the approximate solution of Equation (2) with the initial condition Equation (3) when $n \to \infty$ as follows:

$$\omega(x,\tau) \approx \omega_n(x,\tau) = L^{-1}\{T_n(x,s)\} = \max(e^x - 1, 0) + (e^x - \max(e^x - 1, 0)) \left[\frac{e^{\frac{k\alpha\tau}{k\alpha - k - 1}} + k\alpha - k - 1}{k\alpha - k - 1} \right]. \quad (26)$$

Considering the special case of fractional parameter $\alpha = 1$, we have the exact solution of the problems (2) and (3) as $\omega_{\alpha=1}(x,\tau) = \lim_{n\to\infty,\alpha\to 1} T_n(x,\tau) = e^x \left(1 - e^{-k\tau}\right) + e^{-k\tau} \max(e^x - 1, 0)$.

In Figure 1, the numerical computation of Equation (26) for special case $x = 0.8$ and, in Figure 2, the simulation sketch for $\alpha = 0.35$ in the sense of the Caputo–Fabrizio fractional derivative are presented.

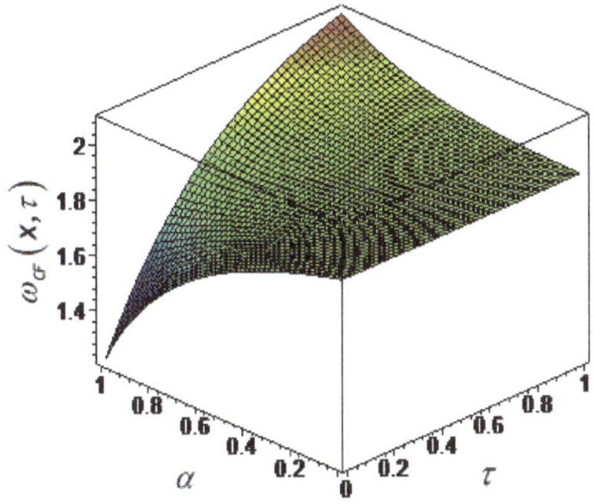

Figure 1. The solution function of Equation (2) in the sense of Caputo–Fabrizio (CF) with respect to $(\alpha, \tau) = [0,1] \times [0,1]$.

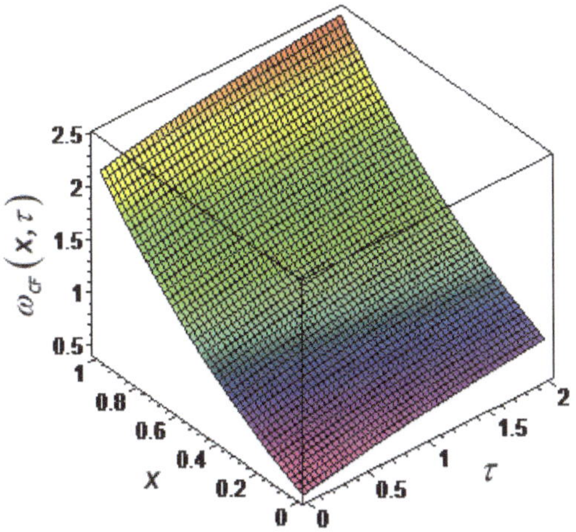

Figure 2. Numerical simulation of Equation (26) in the sense of CF for $\alpha = 0.35$.

Figure 3 shows the European vanilla call option prices, which are given in Equation (1) with exercise price $E = 70$, for fractional values $\alpha = 0.25$, $\alpha = 0.50$, $\alpha = 0.75$ and $\alpha = 1.00$. According to Figure 3, we can say that the option has the lowest price in exercise time of the option $(\tau = T)$, when $\alpha = 1$. As α decreases, the payoff of the option increases. When $\alpha = 0.25$, we observe that the option is overpriced [38].

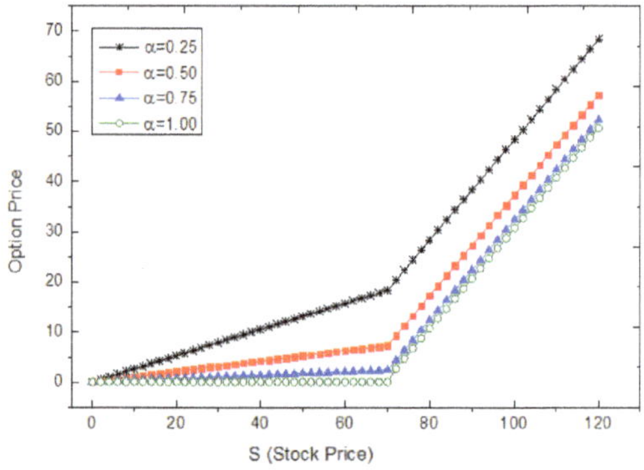

Figure 3. Option prices $V(S,t)$ with respect to underlying asset S for different α values.

3.2. *Fractional Generalized European Option Pricing Problem in the Sense of the Caputo–Fabrizio Derivative*

Secondly, we solve the generalized version of the European vanilla option problems (4) and (5) by using the LHAM constructed in the Caputo–Fabrizio derivative sense. The Laplace transformation of the homogenous term is zero, i.e., $\tilde{v}(x,s) = L\{v(x,\tau)\} = 0$. Now, we obtain the homotopies according to the CF derivative as follows:

$$\begin{aligned}
z^0 &: \Omega_0(x,s) = \tfrac{1}{s}u(x,0) + \left(\tfrac{s+\alpha(1-s)}{s}\right)(0) = \tfrac{\max\left(x-25e^{-0.06},0\right)}{s},\\
z^1 &: \Omega_1(x,s) = -\left(\tfrac{s+\alpha(1-s)}{s}\right)\left[0.08(2+\sin x)^2 \tfrac{\partial^2\Omega_0(x,s)}{\partial x^2} + 0.06x\tfrac{\partial\Omega_0(x,s)}{\partial x} - 0.06\Omega_0(x,s)\right]\\
&= \tfrac{0.06}{s}\left(\tfrac{s+\alpha(1-s)}{s}\right)\left(\max\left(x-25e^{-0.06},0\right)-x\right),\\
z^2 &: \Omega_2(x,s) = -\left(\tfrac{s+\alpha(1-s)}{s}\right)\left[0.08(2+\sin x)^2 \tfrac{\partial^2\Omega_1(x,s)}{\partial x^2} + 0.06x\tfrac{\partial\Omega_1(x,s)}{\partial x} - 0.06\Omega_1(x,s)\right]\\
&= \tfrac{(0.06)^2}{s}\left(\tfrac{s+\alpha(1-s)}{s}\right)^2\left(\max\left(x-25e^{-0.06},0\right)-x\right),\\
&\vdots\\
z^n &: \Omega_n(x,s) = -\left(\tfrac{s+\alpha(1-s)}{s}\right)\left[0.08(2+\sin x)^2 \tfrac{\partial^2\Omega_{n-1}(x,s)}{\partial x^2} + 0.06x\tfrac{\partial\Omega_{n-1}(x,s)}{\partial x} - 0.06\Omega_{n-1}(x,s)\right]\\
&= \tfrac{(0.06)^n}{s}\left(\tfrac{s+\alpha(1-s)}{s}\right)^n\left(\max\left(x-25e^{-0.06},0\right)-x\right),
\end{aligned} \tag{27}$$

By building the n-th order approximate solution, we have:

$$\begin{aligned}
T_n(x,s) &= \textstyle\sum_{j=0}^{n}\Omega_j(x,s)\\
&= \frac{\max\left(x-25e^{-0.06},0\right)}{s} + \frac{\left(\max\left(x-25e^{-0.06},0\right)-x\right)}{s}\sum_{m=1}^{n}\left(0.06\left(\frac{s+\alpha(1-s)}{s}\right)\right)^m
\end{aligned} \tag{28}$$

Applying the inverse LT to Equation (28), when $n \to \infty$, we get the approximate solution of (4) and (5) as follows:

$$\begin{aligned}
\omega(x,\tau) &\approx \omega_n(x,\tau) = L^{-1}\{T_n(x,s)\}\\
&= \max\left(x-25e^{-0.06},0\right) + 16.6667\left(\max\left(x-25e^{-0.06},0\right)-x\right)\frac{\left(e^{\frac{\alpha\tau}{\alpha+15.667}} - 0.06\alpha - 0.94\right)}{\alpha+15.667}.
\end{aligned} \tag{29}$$

For the special case of fractional parameter $\alpha = 1$, we obtain the exact solution of the mentioned problem as $\omega(x,\tau) = \max\left(x-25e^{-0.06},0\right)e^{0.06\tau} + x\left(1-e^{0.06\tau}\right)$, which is the same solution found in [8].

The numerical evaluation of Equation (29) is shown in Figure 4 regarding $x = 1$ in the Caputo–Fabrizio fractional derivative sense. In addition, the numerical simulation of the solution function (29) for different distance values is represented in Figure 5.

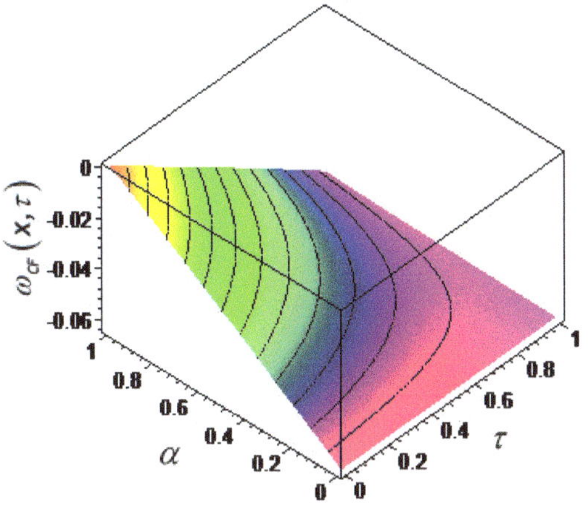

Figure 4. The solution function of (4) in the CF derivative sense with respect to $(\alpha, \tau) = [0,1] \times [0,1]$.

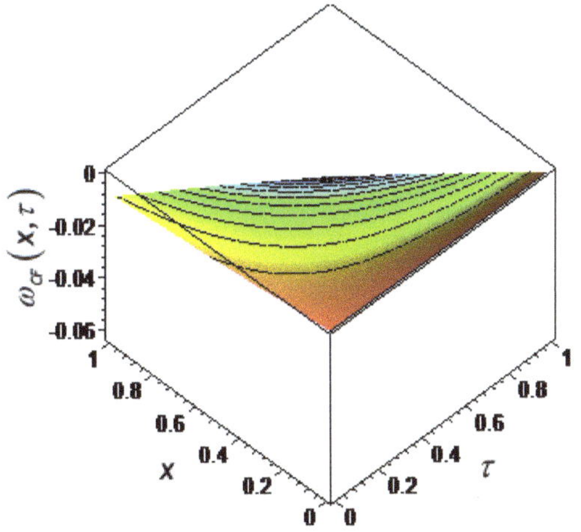

Figure 5. Numerical simulation of (29) in the CF derivative sense for $\alpha = 0.85$.

4. Determining Stabilization and Convergence of Suggested Method

In this part of the study, we have explained the obtained values compatibility test by regarding them as the convergence and the stability of the suggested method. Because the series (25) and (28) converges, these series have to be the solution of initial value problems (2)–(5), respectively. In addition, the solution results represent that the suggested solution technique is convergent and stable. The mentioned method we used in this study provides a good convergence area of the solution. The numerical results found with LHAM are good settlements with the exact solutions. For the purpose of understanding the convergence and stability of the method defined by using the CF operator fractional derivative in Section 2, the amount of the absolute error R_{AE} for some values of x and τ has been presented. In Figure 6, we have also investigated the error rates based on the numerical and exact solution results. According to the results of this stability analysis, it can be concluded that the Caputo–Fabrizio LHAM is an effective and accurate method, for which the series is easily computable in a short time.

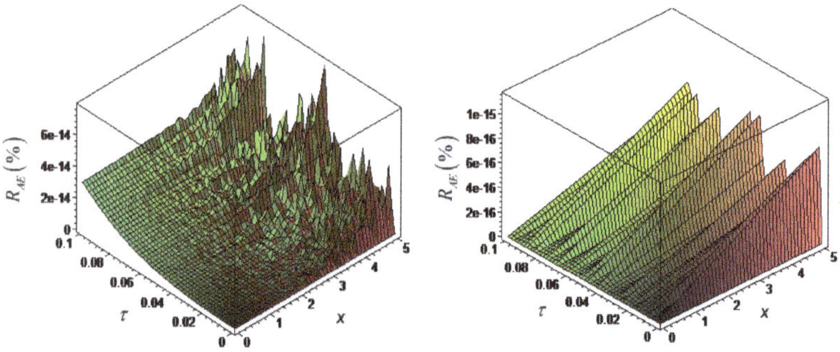

Figure 6. Absolute error rates R_{AE} for some values of x and τ for (2)–(3) (**left**) and (4)–(5) (**right**).

5. Conclusions

In this paper, we have employed approximate-analytical solutions by using a new numerical method, which is described with the Caputo–Fabrizio fractional derivative operator for linear PDEs of time-fractional order. This new fractional operator has a smooth kernel that takes on two different impressions for the spatial and temporal variable. Furthermore, the CF operator has been extremely popular in the last few years. We have demonstrated the efficiencies and accuracies of the suggested method by applying it to the FBS option pricing models with their initial conditions satisfied by the classical European vanilla option. By using the real market values from the finance literature, we can obtain how the option is priced for fractional cases of European call option pricing models. If we consider the European vanilla call option prices, which are given in Equation (1) with exercise price $E = 70$, for special fractional values $\alpha = 0.25$, $\alpha = 0.50$, $\alpha = 0.75$ and $\alpha = 1.00$ in Figure 3, we have concluded that the option has the lowest price in exercise time $(\tau = T)$ of the option, when $\alpha = 1$. Moreover, as α decreases, the payoff of the option increases. When $\alpha = 0.25$, we observe that the option is overpriced [38]. The fractional model suggested in this study can model the price of different financial derivatives like swaps, warrant, etc. The successful applications of the proposed model prove that this model is in complete agreement with the corresponding exact solutions. Besides, in view of their usability, our method is applicable to many initial-boundary problems and fractional linear-nonlinear PDEs. Furthermore, the method is much easier than other homotopy methods, so the LT allows one in many positions to eliminate the inadequacy essentially caused by insufficient conditions, which take part in other approximate-analytical methods like homotopy perturbation method [39].

Acknowledgments: The authors are grateful to the referees for their improvements of the paper.

Author Contributions: Mehmet Yavuz conceived the manuscripts, designed the model and wrote the paper. Necati Özdemir analyzed the solution method and the stabilization; all authors read and approved the final manuscript.

Conflicts of Interest: The authors declare no conflict of interest.

Abbreviations

The following abbreviations are used in this manuscript:

FDE	Fractional differential equation
LHAM	Laplace homotopy analysis method
CF	Caputo–Fabrizio
AB	Atangana–Baleanu
LT	Laplace transform
FC	Fractional calculus
ENSO	El Niño–Southern Oscillation
FBSE	Fractional Black–Scholes equation
GFBSE	Generalized fractional Black–Scholes equation

References

1. Gómez-Aguilar, J.F.; Morales-Delgado, V.F.; Taneco-Hernández, M.A.; Baleanu, D.; Escobar-Jiménez, R.F.; Al Qurashi, M.M. Analytical solutions of the electrical RLC circuit via Liouville–Caputo operators with local and non-local kernels. *Entropy* **2016**, *18*, 402, doi:10.3390/e18080402.
2. Özdemir, N.; Yavuz, M. Numerical solution of fractional Black–Scholes equation by using the multivariate Padé approximation. *Acta Phys. Pol. A.* **2017**, *132*, 1050–1053, doi:10.12693/APhysPolA.132.1050.
3. Yavuz, M.; Özdemir, N.; Okur, Y.Y. Generalized differential transform method for fractional partial differential equation from finance. In Proceedings of the International Conference on Fractional Differentiation and its Applications, Novi Sad, Serbia, 18–20 July 2016; pp. 778–785.
4. Alkahtani, B.S.T.; Koca, I.; Atangana, A. Analysis of a new model of H1N1 spread: Model obtained via Mittag-Leffler function. *Adv. Mech. Eng.* **2017**, *9*, 1–8, doi:10.1177/1687814017705566.
5. Momani, S.; Qaralleh, R. Numerical approximations and Padé approximants for a fractional population growth model. *App. Math. Model.* **2007**, *31*, 1907–1914.
6. Hashim, I.; Abdulaziz, O.; Momani, S. Homotopy analysis method for fractional IVPs. *Commun. Nonlinear Sci. Numer. Simul.* **2009**, *14*, 674–684, doi:10.1016/j.cnsns.2007.09.014.
7. Evirgen, F.; Özdemir, N. Multistage Adomian decomposition method for solving NLP problems over a nonlinear fractional dynamical system. *J. Comput. Nonlinear Dyn.* **2011**, *6*, doi:10.1115/1.4002393.
8. Yavuz, M.; Özdemir, N. A quantitative approach to fractional option pricing problems with decomposition series. *Konuralp J. Math.* **2018**, in press.
9. Elbeleze, A.A.; Kılıçman, A.; Taib, B.M. Homotopy perturbation method for fractional Black–Scholes European option pricing equations using sumudu transform. *Math. Prob. Eng.* **2013**, *2013*, 1–7, doi:10.1155/2013/524852.
10. Yavuz, M. Novel solution methods for initial boundary value problems of fractional order with conformable differentiation. *Int. J. Opt. Control Theor. Appl.* **2018**, *8*, 1–7, doi:10.11121/ijocta.01.2018.00540.
11. Yavuz, M.; Yaşkıran, B. Approximate-analytical solutions of cable equation using conformable fractional operator. *New Trends Math. Sci.* **2017**, *5*, 209–219, doi:10.20852/ntmsci.2017.232.
12. Odibat, Z.; Momani, S.; Erturk, V.S. Generalized differential transform method: Application to differential equations of fractional order. *App. Math. Comput.* **2008**, *197*, 467–477.
13. Chen, W.; Wang, S. A finite difference method for pricing European and American options under a geometric Lévy process. *Management* **2015**, *11*, 241–264.
14. Turut, V.; Güzel, N. On solving partial differential equations of fractional order by using the variational iteration method and multivariate Padé approximations. *Eur. J. Pure Appl. Math.* **2013**, *6*, 147–171.
15. Eroğlu, B.İ.; Avcı, D.; Özdemir, N. Optimal control problem for a conformable fractional heat conduction equation. *Acta Phys. Pol. A.* **2017**, *132*, 658–662, doi:10.12693/APhysPolA.132.658.
16. Evirgen, F. Conformable fractional gradient based dynamic system for constrained optimization problem. *Acta Phys. Pol. A.* **2017**, *132*, 1066–1069, doi:10.12693/APhysPolA.132.1066.
17. Hu, Y.; Øksendal, B.; Sulem, A. Optimal consumption and portfolio in a Black–Scholes market driven by fractional Brownian motion. *Infin. Dimens. Anal. Quantum Prob. Relat. Top.* **2003**, *6*, 519–536.
18. Özdemir, N.; Agrawal, O.P.; Karadeniz, D.; Iskender, B.B. Analysis of an axis-symmetric fractional diffusion-wave problem. *J. Phys. A Math. Theor.* **2009**, *42*, 355208.

19. Caputo, M.; Fabrizio, M. A new definition of fractional derivative without singular kernel. *Prog. Fract. Differ. Appl.* **2015**, *1*, 1–13, doi:10.12785/pfda/010201.
20. Atangana, A.; Alkahtani, B.S.T. New model of groundwater flowing within a confine aquifer: Application of Caputo–Fabrizio derivative. *Arab. J. Geosci.* **2016**, *9*, 1–8, doi:10.1007/s12517-015-2060-8.
21. Singh, J.; Kumar, D.; Nieto, J.J. Analysis of an El Nino-Southern Oscillation model with a new fractional derivative. *Chaos Solitons Fractals* **2017**, *99*, 109–115, doi:10.1016/j.chaos.2017.03.058.
22. Singh, J.; Kumar, D.; Nieto, J.J. Application of Caputo–Fabrizio derivatives to mhd free convection flow of generalized Walters'-B fluid model. *Eur. Phys. J. Plus* **2016**, *131*, 377, doi:10.1140/epjp/i2016-16377-x.
23. Morales-Delgado, V.F.; Gómez-Aguilar, J.F.; Yépez-Martínez, H.; Baleanu, D.; Escobar-Jimenez, R.F.; Olivares-Peregrino, V.H. Laplace homotopy analysis method for solving linear partial differential equations using a fractional derivative with and without kernel singular. *Adv. Differ. Equ.* **2016**, *2016*, 164, doi:10.1186/s13662-016-0891-6.
24. Sheikh, N.A.; Ali, F.; Saqib, M.; Khan, I.; Jan, S.A.A.; Alshomrani, A.S.; Alghamdi, M.S. Comparison and analysis of the Atangana–Baleanu and Caputo–Fabrizio fractional derivatives for generalized Casson fluid model with heat generation and chemical reaction. *Res. Phys.* **2017**, *7*, 789–800, doi:10.1016/j.rinp.2017.01.025.
25. Koca, I.; Atangana, A. Solutions of Cattaneo–Hristov model of elastic heat diffusion with Caputo–Fabrizio and Atangana–Baleanu fractional derivatives. *Therm. Sci.* **2017**, *21*, 2299–2305, doi:10.2298/TSCI160209103K.
26. Hristov, J. Steady-state heat conduction in a medium with spatial non-singular fading memory: Derivation of Caputo–Fabrizio space-fractional derivative with Jeffrey's kernel and analytical solutions. *Therm. Sci.* **2017**, *21*, 827–839, doi:10.2298/TSCI160229115H.
27. Alkahtani, B.S.T.; Atangana, A. Controlling the wave movement on the surface of shallow water with the Caputo–Fabrizio derivative with fractional order. *Chaos Solitons Fractals* **2016**, *89*, 539–546, doi:10.1016/j.chaos.2016.03.012.
28. Alkahtani, B.S.T.; Atangana, A. Analysis of non-homogeneous heat model with new trend of derivative with fractional order. *Chaos Solitons Fractals* **2016**, *89*, 566–571, doi:10.1016/j.chaos.2016.03.027.
29. Hristov, J. Transient heat diffusion with a non-singular fading memory: From the Cattaneo constitutive equation with Jeffrey's kernel to the Caputo–Fabrizio time-fractional derivative. *Therm. Sci.* **2016**, *20*, 757–762, doi:10.2298/TSCI160112019H.
30. Atangana, A.; Koca, I. On the new fractional derivative and application to nonlinear Baggs and Freedman model. *J. Nonlinear Sci. Appl.* **2016**, *9*, 2467–2480.
31. Ghandehari, M.A.M.; Ranjbar, M. European option pricing of fractional Black–Scholes model with new Lagrange multipliers. *Comput. Methods Differ. Equ.* **2014**, *2*, 1–10.
32. Yerlikaya-Özkurt, F.; Vardar-Acar, C.; Yolcu-Okur, Y.; Weber, G.W. Estimation of the Hurst parameter for fractional Brownian motion using the CMARS method. *J. Comput. Appl. Math.* **2014**, *259*, 843–850, doi:10.1016/j.cam.2013.08.001.
33. Black, F.; Scholes, M. The pricing of options and corporate liabilities. *J. Political Econ.* **1973**, *81*, 637–654.
34. Wilmott, P.; Howison, S.; Dewynne, J. *The Mathematics of Financial Derivatives: A Student Introduction*; Cambridge University Press: Cambridge, UK, 1995.
35. Ahmed, E.; Abdusalam, H.A. On modified Black–Scholes equation. *Chaos Solitons Fractals* **2004**, *22*, 583–587, doi:10.1016/j.chaos.2004.02.018.
36. Cen, Z.; Le, A. A robust and accurate finite difference method for a generalized Black–Scholes equation. *J. Comput. Appl. Math.* **2011**, *235*, 3728–3733, doi:10.1016/j.cam.2011.01.018.
37. Meng, L.; Wang, M. Comparison of Black–Scholes formula with fractional Black–Scholes formula in the foreign exchange option market with changing volatility. *Asia-Pac. Finan Markets* **2010**, *17*, 99–111, doi:10.1007/s10690-009-9102-8.
38. Yavuz, M.; Özdemir, N. A different approach to the European option pricing model with new fractional operator. *Math. Model. Nat. Phenom.* **2018**, in press, doi:10.1051/mmnp/2018009.
39. Madani, M.; Fathizadeh, M.; Khan, Y.; Yildirim, A. On the coupling of the homotopy perturbation method and Laplace transformation. *Math. Comput. Model.* **2011**, *53*, 1937–1945, doi:10.1016/j.mcm.2011.01.023.

© 2018 by the authors. Licensee MDPI, Basel, Switzerland. This article is an open access article distributed under the terms and conditions of the Creative Commons Attribution (CC BY) license (http://creativecommons.org/licenses/by/4.0/).

fractal and fractional

MDPI

Article

Fractional Derivatives with the Power-Law and the Mittag–Leffler Kernel Applied to the Nonlinear Baggs–Freedman Model

José Francisco Gómez-Aguilar [1,*] and Abdon Atangana [2]

1 CONACyT-Tecnológico Nacional de México/CENIDET, Interior Internado Palmira S/N, Col. Palmira, C.P. 62490 Cuernavaca, Morelos, Mexico

2 Institute for Groundwater Studies, Faculty of Natural and Agricultural Sciences, University of the Free State, Bloemfontein 9300, South Africa; AtanganaA@ufs.ac.za

* Correspondence: jgomez@cenidet.edu.mx; Tel.: +52-777-362-7770

Received: 20 November 2017; Accepted: 7 February 2018; Published: 9 February 2018

Abstract: This paper considers the Freedman model using the Liouville–Caputo fractional-order derivative and the fractional-order derivative with Mittag–Leffler kernel in the Liouville–Caputo sense. Alternative solutions via Laplace transform, Sumudu–Picard and Adams–Moulton rules were obtained. We prove the uniqueness and existence of the solutions for the alternative model. Numerical simulations for the prediction and interaction between a unilingual and a bilingual population were obtained for different values of the fractional order.

Keywords: Freedman model; Liouville–Caputo derivative; Atangana–Baleanu derivative; fixed point theorem

1. Introduction

Interactions between groups that speak different languages are occurring continuously in several countries in the world due to globalization and cultural openness. Multilingualism is the use of more than one language, either by an individual speaker or by a community of speakers. A mathematical model portraying the interaction dynamics of a population considering bilingual components and a monolingual component was proposed in [1,2]. Baggs in [2] studied the condition under which the bilingual component could persist and conditions under which it could become extinct. The weakness of these models is that they do not take into account the degree of interest in time and also the memory of the interaction, meaning the recall of the original meeting or interaction or contact up to a particular period of time in the present. Fractional calculus is one of the most powerful mathematical tools used in recent decades to model real-world problems in many fields, such as science, technology and engineering. The Liouville–Caputo fractional derivative involves the power-law function. The Liouville–Caputo fractional-order derivative allows usual initial conditions when playing with the integral transform, for instance the Laplace transform [3–5]. Recently, Abdon Atangana and Dumitru Baleanu proposed two fractional-order operators involving the generalized Mittag–Leffler function. The generalized Mittag–Leffler function was introduced in the literature to improve the limitations posed by the power-law [6–12]. The two-parametric, three-parametric, four-parametric and multiple Mittag–Leffler functions were presented by Wiman, Prabhakar, Shukla and Srivastava in [13–18]. The kernel used in Atangana–Baleanu fractional differentiation appears naturally in several physical problems as generalized exponential decay and as a power-law asymptotic for a very large time [19–24]. The choice of this derivative is motivated by the fact that the interaction is not local, but global, and also, the trend observed in the field does not follow the power-law. The generalized

Mittag–Leffler function completely induced the effect of memory, which is very important in the nonlinear Baggs–Freedman model.

Atangana and Koca in [25] studied the nonlinear Baggs and Freedman model. Starting from the integer-order Freedman model presented by [25], we have:

$$\begin{aligned} \mathcal{D}_t x_1 &= (A_1 - M_1 - F_1)x_1(t) - L_1 x_1^2(t) - \alpha \cdot \tfrac{x_1(t)x_2(t)}{1+x_1(t)} + G_1 A_2 x_2(t), \\ \mathcal{D}_t x_2 &= (A_2 - M_2 - F_2)x_2(t) - L_2 x_2^2(t) + \alpha \cdot \tfrac{x_1(t)x_2(t)}{1+x_1(t)} - G_1 A_2 x_2(t), \end{aligned} \tag{1}$$

where $0 < A_i, M_i, |F_i| \leq 1$ are the birth, death and emigration parameters for $i \in [1,2]$. $0 < G_1 \leq 1$ is the infant language acquisition parameter, that proportion of births in the x_2 population raised unilingually. $0 < \alpha \leq 1$ is the non-infant language acquisition rate, the proportion of x_1 learning the x_2 language per unit time after infancy. The term $\alpha \cdot \frac{x_1(t)x_2(t)}{1+x_1(t)}$ describes that part of population x_1 lost to x_2 due to virtual predation on the part of x_2 [26].

The aim of this work is to obtain alternative representations of the Freedman model considering Liouville–Caputo and Atangana–Baleanu–Caputo fractional derivatives. The paper is organized as follows: Section 2 introduces the fractional operators. Alternative representations of the Freedman model are shown in Section 3. Finally, in Section 4, we conclude the manuscript.

2. Fractional Operators

The Liouville–Caputo fractional-order derivative of order γ is defined by [27]:

$${}_a^C\mathcal{D}_t^\gamma\{f(t)\} = \frac{1}{\Gamma(1-\gamma)}\int_a^t \dot{f}(\theta)(t-\theta)^{-\gamma}d\theta, \qquad 0 < \gamma \leq 1. \tag{2}$$

The Laplace transform to Liouville–Caputo fractional-order derivative gives [27]:

$$\mathcal{L}[{}_a^C\mathcal{D}_t^\gamma f(t)] = S^\gamma F(S) - \sum_{k=0}^{m-1} S^{\gamma-k-1} f^{(k)}(0). \tag{3}$$

The Atangana–Baleanu–Caputo fractional-order derivative is defined as follows [19–22,24]:

$${}_a^{ABC}\mathcal{D}_t^\gamma\{f(t)\} = \frac{B(\gamma)}{1-\gamma}\int_a^t \dot{f}(\theta)E_\gamma\Big[-\frac{\gamma}{1-\gamma}(t-\theta)^\gamma\Big]d\theta, \qquad 0 < \gamma \leq 1, \tag{4}$$

where $B(\beta) = B(0) = B(1) = 1$ is a normalization function and E_γ is the Mittag–Leffler function [6–12]. The Mittag–Leffler kernel is a combination of both the exponential-law and power-law. For this fractional derivative, we have at the same time the power-law and the stretched exponential as the waiting time distribution.

The Laplace transform of Equation (4) is defined as follows:

$$\begin{aligned} \mathcal{L}[{}_a^{ABC}\mathcal{D}_t^\gamma f(t)](s) &= \tfrac{B(\gamma)}{1-\gamma}\mathcal{L}\Big[\int_a^t \dot{f}(\theta)E_\gamma\Big[-\gamma\tfrac{(t-\theta)^\gamma}{1-\gamma}\Big]d\theta\Big] \\ &= \tfrac{B(\gamma)}{1-\gamma}\tfrac{s^\gamma\mathcal{L}[f(t)](s) - s^{\gamma-1}f(0)}{s^\gamma + \frac{\gamma}{1-\gamma}}. \end{aligned} \tag{5}$$

The Sumudu transform is derived from the classical Fourier integral [28]. The Sumudu transform of Equation (4) is defined as:

$$ST\{{}_a^{ABC}\mathcal{D}_t^\gamma f(t)\} = \frac{B(\gamma)}{1-\gamma}\Big(\gamma\Gamma(\gamma+1)E_\gamma\Big(-\frac{1}{1-\gamma}u^\gamma\Big)\Big) \times [ST(f(t)) - f(0)]. \tag{6}$$

The Atangana–Baleanu fractional integral of order γ of a function $f(t)$ is defined as:

$$ {}^{AB}_{a}I^{\gamma}_{t}f(t) = \frac{1-\gamma}{B(\gamma)}f(t) + \frac{\gamma}{B(\gamma)\Gamma(\gamma)}\int_0^t f(s)(t-s)^{\gamma-1}ds. \tag{7} $$

3. Freedman Model

In this section, we obtain alternative representations of the Freedman model considering the Liouville–Caputo fractional derivative, and the special solution is obtained using a Laplace transform method.

3.1. Freedman Model with the Power-Law Kernel

Considering Equation (2), the modified Freedman model with the power-law kernel is given as:

$$ \begin{aligned} {}^{C}_{0}\mathcal{D}^{\gamma}_{t}x_1(t) &= (A_1-M_1-F_1)x_1(t) - L_1x_1^2(t) - \alpha\cdot\tfrac{x_1(t)x_2(t)}{1+x_1(t)} + G_1A_2x_2(t), \\ {}^{C}_{0}\mathcal{D}^{\gamma}_{t}x_2(t) &= (A_2-M_2-F_2)x_2(t) - L_2x_2^2(t) + \alpha\cdot\tfrac{x_1(t)x_2(t)}{1+x_1(t)} - G_1A_2x_2(t), \end{aligned} \tag{8} $$

where $0<\gamma\le 1$ is the fractional order, $x_1(t)$ represents the interaction of the majority unilingual population and $x_2(t)$ is a bilingual population.

Applying the Laplace transform operator (3) and the inverse Laplace transform on both sides of Equation (8), we obtain:

$$ \begin{aligned} x_1(t) &= x_1(0) + \mathscr{L}^{-1}\Big\{\tfrac{1}{s^{\gamma}}\mathscr{L}\Big[(A_1-M_1-F_1)x_1(t) - L_1x_1^2(t) - \alpha\tfrac{x_1(t)x_2(t)}{1+x_1(t)} + G_1A_2x_2(t)\Big](s)\Big\}(t), \\ x_2(t) &= x_2(0) + \mathscr{L}^{-1}\Big\{\tfrac{1}{s^{\gamma}}\mathscr{L}\Big[(A_2-M_2-F_2)x_2(t) - L_2x_2^2(t) + \alpha\tfrac{x_1(t)x_2(t)}{1+x_1(t)} - G_1A_2x_2(t)\Big](s)\Big\}(t). \end{aligned} \tag{9} $$

The following iterative formula is then proposed:

$$ \begin{aligned} x_{1(n)}(t) &= \mathscr{L}^{-1}\Big\{\tfrac{1}{s^{\gamma}}\mathscr{L}\Big[(A_1-M_1-F_1)x_{1(n-1)}(t) - L_1x^2_{1(n-1)}(t) \\ &\quad -\alpha\tfrac{x_{1(n-1)}(t)x_{2(n-1)}(t)}{1+x_{1(n-1)}(t)} + G_1A_2x_{2(n-1)}(t)\Big](s)\Big\}(t), \\ x_{2(n)}(t) &= \mathscr{L}^{-1}\Big\{\tfrac{1}{s^{\gamma}}\mathscr{L}\Big[(A_2-M_2-F_2)x_{2(n-1)}(t) - L_2x^2_{1(n-1)}(t) \\ &\quad +\alpha\tfrac{x_{1(n-1)}(t)x_{2(n-1)}(t)}{1+x_{1(n-1)}(t)} - G_1A_2x_{2(n-1)}(t)\Big](s)\Big\}(t), \end{aligned} \tag{10} $$

where,

$$ x_{1(0)}(t) = x_1(0); \qquad x_{2(0)}(t) = x_2(0), \tag{11} $$

where the approximate solution is assumed to be obtained as a limit when "n" tend to infinity:

$$ x_1(t) = \lim_{n\to\infty} x_{1(n)}(t); \qquad x_2(t) = \lim_{n\to\infty} x_{2(n)}(t). \tag{12} $$

3.2. Stability Analysis of the Iteration Method

Theorem 1. *We demonstrate that the recursive method given by Equation* (10) *is stable.*

Proof. It is possible to find two positive constants Y and Z such that, for all:

$$ 0\le t\le T\le\infty, \qquad ||x_1(t)||<Y \qquad \text{and} \qquad ||x_1(t)||<Z. \tag{13} $$

Now, we consider a subset of $C_2((a,b)(0,T))$ defined by:

$$ H = \Big\{\varphi : (a,b)(0,T)\to H, \qquad \frac{1}{\Gamma(\gamma)}\int (t-\varphi)^{\gamma-1}v(\varphi)u(\varphi)d\varphi < \infty\Big\}, \tag{14} $$

we now consider the operator ϕ defined as:

$$\phi(x_1, x_2) = \begin{cases} (A_1 - M_1 - F_1)x_1(t) - L_1x_1^2(t) - \alpha \cdot \frac{x_1(t)x_2(t)}{1+x_1(t)} + G_1A_2x_2(t), \\ (A_2 - M_2 - F_2)x_2(t) - L_2x_2^2(t) + \alpha \cdot \frac{x_1(t)x_2(t)}{1+x_1(t)} - G_1A_2x_2(t). \end{cases}$$

Then:

$$= \begin{cases} < \phi(x_1, x_2) - \phi(X_1, X_2), \phi(x_1 - X_1, x_2 - X_2) >, \\ < (A_1 - M_1 - F_1)(x_1(t) - X_1(t)) - L_1(x_1(t) - X_1(t))^2 \\ -\alpha \cdot \frac{(x_1(t)-X_1(t))(x_2(t)-X_2(t))}{1+(x_1(t)-X_1(t))} + G_1A_2(x_2(t) - X_2(t)) >, \\ < (A_2 - M_2 - F_2)(x_2(t) - X_2(t)) - L_2(x_1(t) - X_2(t))^2 \\ +\alpha \cdot \frac{(x_1(t)-X_1(t))(x_2(t)-X_2(t))}{1+(x_1(t)-X_1(t))} - G_1A_2(x_2(t) - X_2(t)) >, \end{cases}$$

where,

$$x_1(t) \neq X_1(t), \qquad \text{and} \qquad x_2(t) \neq X_2(t). \tag{15}$$

Applying the absolute value on both sides, we have:

$$\begin{aligned} &< \phi(x_1, x_2) - \phi(X_1, X_2), \phi(x_1 - X_1, x_2 - X_2) >, \\ &= \begin{cases} \Big\{(A_1 - M_1 - F_1) - L_1||x_1(t) - X_1(t)|| \\ -\alpha \cdot \frac{||x_2(t)-X_2(t)||}{1+||x_1(t)-X_1(t)||^2} + G_1A_2\frac{||x_2(t)-X_2(t)||}{||x_1(t)-X_1(t)||}\Big\}||x_1(t) - X_1(t)||^2, \\ \Big\{(A_2 - M_2 - F_2) - L_2||x_2(t) - X_2(t)|| \\ +\alpha \cdot \frac{||x_1(t)-X_1(t)||}{1+||x_1(t)-X_1(t)||^2} - G_1A_2\Big\}||x_2(t) - X_2(t)||^2. \end{cases} \end{aligned} \tag{16}$$

Then,

$$\begin{aligned} &< \phi(x_1, x_2) - \phi(X_1, X_2), \phi(x_1 - X_1, x_2 - X_2) >, \\ &< \begin{cases} \Big\{(A_1 - M_1 - F_1) + L_1||x_1(t) - X_1(t)|| \\ +\alpha \cdot \frac{||x_2(t)-X_2(t)||}{1+||x_1(t)-X_1(t)||^2} + G_1A_2\frac{||x_2(t)-X_2(t)||}{||x_1(t)-X_1(t)||}\Big\}||x_1(t) - X_1(t)||^2, \\ \Big\{(A_2 - M_2 - F_2) + L_2||x_2(t) - X_2(t)|| \\ +\alpha \cdot \frac{||x_1(t)-X_1(t)||}{1+||x_1(t)-X_1(t)||^2} + G_1A_2\Big\}||x_2(t) - X_2(t)||^2, \end{cases} \end{aligned} \tag{17}$$

where,

$$\begin{aligned} &< \phi(x_1, x_2) - \phi(X_1, X_2), \phi(x_1 - X_1, x_2 - X_2) >, \\ &< \begin{cases} M||x_1(t) - X_1(t)||^2, \\ N||x_2(t) - X_2(t)||^2, \end{cases} \end{aligned} \tag{18}$$

with:

$$M = (A_1 - M_1 - F_1) + L_1||x_1(t) - X_1(t)|| + \alpha \cdot \frac{||x_2(t) - X_2(t)||}{1 + ||x_1(t) - X_1(t)||^2} + G_1A_2\frac{||x_2(t) - X_2(t)||}{||x_1(t) - X_1(t)||},$$

and:

$$N = (A_2 - M_2 - F_2) + L_2||x_2(t) - X_2(t)|| + \alpha \cdot \frac{||x_1(t) - X_1(t)||}{1 + ||x_1(t) - X_1(t)||^2} + G_1A_2. \tag{19}$$

Furthermore, if we consider a given non-null vector (x_1, x_2), then using the some routine as the above case, we obtain:

$$\begin{aligned} &< \phi(x_1, x_2) - \phi(X_1, X_2), \phi(x_1 - X_1, x_2 - X_2) >, \\ &< \begin{cases} M||x_1(t) - X_1(t)||||x_1(t)||, \\ N||x_2(t) - X_2(t)||||x_2(t)||, \end{cases} \end{aligned} \tag{20}$$

From the results obtained in Equations (18) and (20), we conclude that the iterative method used is stable. This complete the proof. □

Now, we consider the Adams method [29,30] to solve the system given by Equation (8). The basic idea of the n-step Adams–Bashforth method is to use a polynomial interpolation for $f(t, y(t))$ passing through n points: (t_i, f_i), (t_{i-1}, f_{i-1}), ..., (t_{i-n+1}, f_{i-n+1}). Correspondingly, the n-step Adams–Moulton method uses a polynomial interpolation for $f(t, y(t))$ passing through $n+1$ points: (t_{i+1}, f_{i+1}), (t_i, f_i), ..., (t_{i-n}, f_{i-n}).

The fractional Adams method is derived as follows [30]:

$$\begin{aligned} f^P_{k+1} &= \sum_{j=0}^{n-1} \frac{t^j_{k+1}}{j!} f_0^{(j)} + \frac{1}{\Gamma(\gamma)} \sum_{j=0}^{k} b_{j,k+1} g(t_j, f_j), \\ f_{k+1} &= \sum_{j=0}^{n-1} \frac{t^j_{k+1}}{j!} f_0^{(j)} + \frac{1}{\Gamma(\gamma)} \left(\sum_{j=0}^{k} a_{j,k+1} g(t_j, f_j) + a_{k+1,k+1} g(t_{k+1}, f^P_{k+1}) \right), \end{aligned} \tag{21}$$

where,

$$\begin{aligned} a_{j,k+1} &= \frac{h^\gamma}{\gamma(\gamma+1)} \cdot \begin{cases} (k^{\gamma+1} - (k-\gamma)(k+1)^\gamma) & j = 0, \\ ((k-j+2)^{\gamma+1} + (k-j)^{\gamma+1} - 2(k-j+1)^{\gamma+1}) & 1 \le j \le k, \\ 1 & j = k+1, \end{cases} \\ b_{j,k+1} &= \frac{h^\gamma}{\gamma}((k+1-j)^\gamma - (k-j)^\gamma), \qquad j = 0, 1, 2, ..., k. \end{aligned} \tag{22}$$

Following this procedure, we can propose a numerical solution for System (8) using the Adams method (21) as follows:

$$\begin{aligned} x_1(t) &= \sum_{k=0}^{n-1} x_1(0)^{(k)} \frac{t^k}{k!} + \frac{1}{\Gamma(\gamma)} \int_0^t (t-u)^{\gamma-1} \Big[(A_1 - M_1 - F_1) x_1(u) - L_1 x_1^2(u) - \alpha \cdot \tfrac{x_1(u) x_2(u)}{1 + x_1(u)} + G_1 A_2 x_2(u) \Big] du, \\ x_2(t) &= \sum_{k=0}^{n-1} x_2(0)^{(k)} \frac{t^k}{k!} + \frac{1}{\Gamma(\gamma)} \int_0^t (t-u)^{\gamma-1} \Big[(A_2 - M_2 - F_2) x_2(u) - L_2 x_2^2(u) + \alpha \cdot \tfrac{x_1(u) x_2(u)}{1 + x_1(u)} - G_1 A_2 x_2(u) \Big] du. \end{aligned} \tag{23}$$

3.3. *Freedman Model with the Mittag–Leffler Kernel*

Considering Equation (4), the modified Freedman model is given as:

$$\begin{aligned} {}^{ABC}_{0}\mathcal{D}^\gamma_t x_1(t) &= (A_1 - M_1 - F_1) x_1(t) - L_1 x_1^2(t) - \alpha \cdot \tfrac{x_1(t) x_2(t)}{1 + x_1(t)} + G_1 A_2 x_2(t), \\ {}^{ABC}_{0}\mathcal{D}^\gamma_t x_2(t) &= (A_2 - M_2 - F_2) x_2(t) - L_2 x_2^2(t) + \alpha \cdot \tfrac{x_1(t) x_2(t)}{1 + x_1(t)} - G_1 A_2 x_2(t), \end{aligned} \tag{24}$$

where $0 < \gamma \le 1$ is the fractional order, $x_1(t)$ represents the interaction of the majority unilingual population and $x_2(t)$ is a bilingual population.

Now, we obtain an alternative solution using an iterative scheme. The technique involves coupling the Sumudu transform and its inverse. The Sumudu transform is an integral transform similar to the Laplace transform, introduced by Watugala to solve differential equations [28–32]. Applying the Sumudu transform (6) and the inverse Sumudu transform on both sides of the system (24) yields:

$$\begin{aligned} x_1(t) &= x_1(0) + ST^{-1} \Big\{ \frac{1-\gamma}{B(\gamma)\gamma\Gamma(\gamma+1) E_\gamma \left(-\frac{1}{1-\gamma} u^\gamma \right)} \cdot ST \Big[(A_1 - M_1 - F_1) x_1(t) - L_1 x_1^2(t) - \alpha \tfrac{x_1(t) x_2(t)}{1 + x_1(t)} + G_1 A_2 x_2(t) \Big] \Big\}, \\ x_2(t) &= x_2(0) + ST^{-1} \Big\{ \frac{1-\gamma}{B(\gamma)\gamma\Gamma(\gamma+1) E_\gamma \left(-\frac{1}{1-\gamma} u^\gamma \right)} \cdot ST \Big[(A_2 - M_2 - F_2) x_2(t) - L_2 x_2^2(t) + \alpha \tfrac{x_1(t) x_2(t)}{1 + x_1(t)} - G_1 A_2 x_2(t) \Big] \Big\}. \end{aligned} \tag{25}$$

The following recursive formula for Equation (25) is obtained:

$$\begin{aligned} x_{1(n+1)}(t) = & \; x_{1(n)}(0) + ST^{-1}\Big\{\frac{1-\gamma}{B(\gamma)\gamma\Gamma(\gamma+1)E_\gamma\left(-\frac{1}{1-\gamma}u^\gamma\right)} \\ & \cdot ST\Big[(A_1 - M_1 - F_1)x_{1(n)}(t) - L_1 x^2_{1(n)}(t) \\ & -\alpha\frac{x_{1(n)}(t)x_{2(n)}(t)}{1+x_{1(n)}(t)} + G_1 A_2 x_{2(n)}(t)\Big]\Big\}, \\ x_{2(n+1)}(t) = & \; x_{1(n)}(0) + ST^{-1}\Big\{\frac{1-\gamma}{B(\gamma)\gamma\Gamma(\gamma+1)E_\gamma\left(-\frac{1}{1-\gamma}u^\gamma\right)} \\ & \cdot ST\Big[(A_2 - M_2 - F_2)x_{2(n)}(t) - L_2 x^2_{2(n)}(t) \\ & +\alpha\frac{x_{1(n)}(t)x_{2(n)}(t)}{1+x_{1(n)}(t)} - G_1 A_2 x_{2(n)}(t)\Big]\Big\}, \end{aligned} \tag{26}$$

and the solution of Equation (26) is provided by:

$$x_1(t) = \lim_{n\to\infty} x_{1(n)}(t); \qquad x_2(t) = \lim_{n\to\infty} x_{2(n)}(t). \tag{27}$$

3.4. Stability Analysis of the Iteration Method

Now, we provide in detail the stability analysis of this method and show the uniqueness of the special solutions using the fixed point theory and properties of the inner product and the Hilbert space, respectively.

Let $(X, |\cdot|)$ be a Banach space and Ha self-map of X. Let $z_{n+1} = g(H, z_n)$ be a particular recursive procedure. The following conditions must be satisfied for $z_{n+1} = Hz_n$.

1. The fixed point set of H has at least one element.
2. z_n converges to a point $P \in F(H)$.
3. $\lim_{n\to\infty} x_n(t) = P$.

Property 1. *Let $(X, |\cdot|)$ be a Banach space and Ha self-map of X satisfying:*

$$||H_x - H_z|| \leq \eta||X - H_x|| + \eta||x - z||, \tag{28}$$

for all $x, z \in X$, where $0 \leq \eta, 0 \leq \eta < 1$. Suppose that H is Picard H-stable.

Considering the following recursive formula, we have:

$$\begin{aligned} x_{1(n+1)}(t) = & \; x_{1(n)}(0) + S^{-1}\Big\{\frac{1-\gamma}{B(\gamma)\gamma\Gamma(\gamma+1)E_\gamma\left(-\frac{1}{1-\gamma}u^\gamma\right)} \\ & \cdot S\Big[(A_1 - M_1 - F_1)x_{1(n)}(t) - L_1 x^2_{1(n)}(t) \\ & -\alpha\frac{x_{1(n)}(t)x_{2(n)}(t)}{1+x_{1(n)}(t)} + G_1 A_2 x_{2(n)}(t)\Big]\Big\}, \\ x_{2(n+1)}(t) = & \; x_{2(n)}(0) + S^{-1}\Big\{\frac{1-\gamma}{B(\gamma)\gamma\Gamma(\gamma+1)E_\gamma\left(-\frac{1}{1-\gamma}u^\gamma\right)} \\ & \cdot S\Big[(A_2 - M_2 - F_2)x_{2(n)}(t) - L_2 x^2_{2(n)}(t) \\ & +\alpha\frac{x_{1(n)}(t)x_{2(n)}(t)}{1+x_{1(n)}(t)} - G_1 A_2 x_{2(n)}(t)\Big]\Big\}, \end{aligned} \tag{29}$$

where:

$$\frac{1-\gamma}{B(\gamma)\gamma\Gamma(\gamma+1)E_\gamma\left(-\frac{1}{1-\gamma}u^\gamma\right)}, \tag{30}$$

correspond to the fractional Lagrange multiplier.

Theorem 2. *Let K be a self-map defined as:*

$$\begin{aligned}
K[x_{1(n+1)}(t)] = & \; x_{1(n+1)}(t) = x_{1(n)}(t) + S^{-1}\Big\{\frac{1-\gamma}{B(\gamma)\gamma\Gamma(\gamma+1)E_{\gamma}\left(-\frac{1}{1-\gamma}u^{\gamma}\right)} \\
& \cdot S\Big[(A_1 - M_1 - F_1)x_{1(n)}(t) - L_1 x^2_{1(n)}(t) - \alpha\frac{x_{1(n)}(t)x_{2(n)}(t)}{1+x_{1(n)}(t)} + G_1 A_2 x_{2(n)}(t)\Big]\Big\}, \\
K[x_{2(n+1)}(t)] = & \; x_{2(n+1)}(t) = x_{2(n)}(t) + S^{-1}\Big\{\frac{1-\gamma}{B(\gamma)\gamma\Gamma(\gamma+1)E_{\gamma}\left(-\frac{1}{1-\gamma}u^{\gamma}\right)} \\
& \cdot S\Big[(A_2 - M_2 - F_2)x_{2(n)}(t) - L_2 x^2_{2(n)}(t) + \alpha\frac{x_{1(n)}(t)x_{2(n)}(t)}{1+x_{1(n)}(t)} - G_1 A_2 x_{2(n)}(t)\Big]\Big\},
\end{aligned} \tag{31}$$

is K-stable in $L^1(a,b)$ *if:*

$$\begin{aligned}
1 + (A_1 - M_1 - F_1)f(\gamma) - L_1||x_{1(n)}(t) + x_{1(m)}(t)||d(\gamma) - \alpha\frac{\epsilon\beta+\epsilon+\theta}{\eta\rho}\omega(\gamma) + G_1A_2k(\gamma) < 1, \\
1 + (A_2 - M_2 - F_2)i(\gamma) - L_2||x_{2(n)}(t) + x_{2(m)}(t)||r(\gamma) + \alpha\frac{\epsilon\beta+\epsilon+\theta}{\eta\rho}s(\gamma) - G_1A_2o(\gamma) < 1,
\end{aligned} \tag{32}$$

where $f(\gamma), d(\gamma), \omega(\gamma), k(\gamma), i(\gamma), r(\gamma), s(\gamma)$ *and* $o(\gamma)$ *are functions from:*

$$S^{-1}\left\{\frac{1-\gamma}{B(\gamma)\gamma\Gamma(\gamma+1)E_{\gamma}\left(-\frac{1}{1-\gamma}u^{\gamma}\right)}S\right\}. \tag{33}$$

Proof. The proof consists of showing that K has a fixed point. To achieve this, we consider:

$$\begin{aligned}
& K[x_{1(n+1)}(t)] - K[x_{1(m+1)}(t)] = x_{1(n)}(0) - x_{1(m)}(0) + S^{-1}\left\{\frac{1-\gamma}{B(\gamma)\gamma\Gamma(\gamma+1)E_{\gamma}\left(-\frac{1}{1-\gamma}u^{\gamma}\right)}\right. \\
& \cdot S\Big[((A_1 - M_1 - F_1)x_{1(n)}(t) - (A_1 - M_1 - F_1)x_{1(m)}(t)) - (L_1x^2_{1(n)}(t) + L_1x^2_{1(m)}(t)) \\
& \left. - \left(\alpha\frac{(x_{1(n)}(t) - x_{1(m)}(t))\cdot(x_{2(n)}(t) - x_{2(m)}(t))}{1 + (x_{1(n)}(t) - x_{1(m)}(t)}\right) + G_1A_2(x_{2(n)}(t) - x_{2(m)}(t))\Big]\right\},
\end{aligned}$$

and:

$$\begin{aligned}
& K[x_{2(n+1)}(t)] - K[x_{2(m+1)}(t)] = x_{2(n)}(0) - x_{2(m)}(0) + S^{-1}\left\{\frac{1-\gamma}{B(\gamma)\gamma\Gamma(\gamma+1)E_{\gamma}\left(-\frac{1}{1-\gamma}u^{\gamma}\right)}\right. \\
& \cdot S\Big[((A_2 - M_2 - F_2)x_{2(n)}(t) - (A_2 - M_2 - F_2)x_{2(m)}(t)) - (L_2x^2_{2(n)}(t) + L_2x^2_{2(m)}(t)) \\
& \left. + \left(\alpha\frac{(x_{1(n)}(t) - x_{1(m)}(t))\cdot(x_{2(n)}(t) - x_{2(m)}(t))}{1 + (x_{1(n)}(t) - x_{1(m)}(t)}\right) - G_1A_2(x_{2(n)}(t) - x_{2(m)}(t))\Big]\right\}.
\end{aligned} \tag{34}$$

Using the properties of the norm and considering the triangular inequality, we get:

$$\begin{aligned}
& ||K[x_{1(n)}(t)] - K[x_{1(m)}(t)]|| \leq ||x_{1(n)}(t) - x_{1(m)}(t)|| + S^{-1}\left\{\frac{1-\gamma}{B(\gamma)\gamma\Gamma(\gamma+1)E_{\gamma}\left(-\frac{1}{1-\gamma}u^{\gamma}\right)}\right. \\
& \cdot S\Big[((A_1 - M_1 - F_1)x_{1(n)}(t) - (A_1 - M_1 - F_1)x_{1(m)}(t)) - (L_1x^2_{1(n)}(t) + L_1x^2_{1(m)}(t)) \\
& \left. + \left(\alpha\frac{(x_{1(n)}(t) - x_{1(m)}(t))\cdot(x_{2(n)}(t) - x_{2(m)}(t))}{1 + (x_{1(n)}(t) - x_{1(m)}(t)}\right) - G_1A_2(x_{2(n)}(t) - x_{2(m)}(t))\Big]\right\}.
\end{aligned} \tag{35}$$

we consider that the solutions play the some role, i.e., $||x_{2(n)}(t) - x_{2(m)}(t)|| \cong ||x_{1(n)}(t) - x_{1(m)}(t)||$.

Using the linearity of the inverse Sumudu transform, we get:

$$\begin{aligned}
&||K[x_{1(n)}(t)] - K[x_{1(m)}(t)]|| \leq ||x_{1(n)}(t) - x_{1(m)}(t)|| \\
&+S^{-1}\left\{\frac{1-\gamma}{B(\gamma)\gamma\Gamma(\gamma+1)E_\gamma\left(-\frac{1}{1-\gamma}u^\gamma\right)}S\Big(||(A_1 - M_1 - F_1)[x_{1(n)}(t) - x_{1(m)}(t)]||\Big)\right\} \\
&+S^{-1}\left\{\frac{1-\gamma}{B(\gamma)\gamma\Gamma(\gamma+1)E_\gamma\left(-\frac{1}{1-\gamma}u^\gamma\right)}S\Big(|| - L_1[x_{1(n)}^2(t) - x_{1(m)}^2(t)]||\Big)\right\} \\
&+S^{-1}\left\{\frac{1-\gamma}{B(\gamma)\gamma\Gamma(\gamma+1)E_\gamma\left(-\frac{1}{1-\gamma}u^\gamma\right)}\right. \\
&\cdot S\left(\left|\left| - \alpha\frac{(x_{1(m)}(t)(x_{1(n)}(t))((x_{1(n)}(t) - x_{1(m)}(t))}{1+(x_{1(n)}(t)(1+x_{1(m)}(t))} + \frac{x_{1(n)}(t)(x_{1(n)}(t) - x_{1(m)}(t)) + x_{2(n)}(t)(x_{2(n)}(t) - x_{1(m)}(t))}{1+(x_{1(n)}(t)(1+x_{1(m)}(t))}\right|\right|\right)\Bigg\} \\
&+S^{-1}\left\{\frac{1-\gamma}{B(\gamma)\gamma\Gamma(\gamma+1)E_\gamma\left(-\frac{1}{1-\gamma}u^\gamma\right)}S||G_1A_2[x_{1(n)}(t) - x_{1(m)}(t))]||\right\}.
\end{aligned} \tag{36}$$

Since $x_{1(n)}(t)$ and $x_{1(m)}(t)$ are bounded, we can find the following positive constants, ϵ, β, θ, η and ρ such that for all t:

$$\begin{aligned}
&||x_{1(n)}(t)|| \leq \epsilon, \qquad ||x_{1(m)}(t)|| \leq \beta, \qquad ||1 + x_{1(n)}(t)|| \leq \rho, \\
&||x_{2(n)}(t)|| \leq \theta, \qquad ||1 + x_{1(m)}(t)|| \leq \eta, \qquad (n,m) \in N \times N.
\end{aligned} \tag{37}$$

Considering Equations (36) and (37), we obtain:

$$\begin{aligned}
&||K[x_{1(n)}(t)] - K[x_{1(m)}(t)]|| \leq ||x_{1(n)}(t) - x_{1(m)}(t)|| \\
&\cdot\left(1 + (A_1 - M_1 - F_1)f(\gamma) - L_1||x_{1(n)}(t) + x_{1(m)}(t)||d(\gamma) - \alpha\frac{\epsilon\beta+\epsilon+\theta}{\eta\rho}\omega(\gamma) + G_1A_2k(\gamma)\right),
\end{aligned} \tag{38}$$

and:

$$\begin{aligned}
&||K[x_{2(n)}(t)] - K[x_{2(m)}(t)]|| \leq ||x_{2(n)}(t) - x_{2(m)}(t)|| \\
&\cdot\left(1 + (A_2 - M_2 - F_2)i(\gamma) - L_2||x_{2(n)}(t) + x_{2(m)}(t)||r(\gamma) + \alpha\frac{\epsilon\beta+\epsilon+\theta}{\eta\rho}s(\gamma) - G_1A_2o(\gamma)\right),
\end{aligned} \tag{39}$$

where $f(\gamma), d(\gamma), \omega(\gamma), k(\gamma), i(\gamma), r(\gamma), s(\gamma)$ and $o(\gamma)$ are functions from (33).

We next show that K satisfies Property 1. Consider Equations (38) and (39), yielding:

$$\eta(0,0), \eta = \begin{cases} 1 + (A_1 - M_1 - F_1)f(\gamma) - L_1||x_{1(n)}(t) + x_{1(m)}(t)||d(\gamma) - \alpha\frac{\epsilon\beta+\epsilon+\theta}{\eta\rho}\omega(\gamma) + G_1A_2k(\gamma), \\ 1 + (A_2 - M_2 - F_2)i(\gamma) - L_2||x_{2(n)}(t) + x_{2(m)}(t)||r(\gamma) + \alpha\frac{\epsilon\beta+\epsilon+\theta}{\eta\rho}s(\gamma) - G_1A_2o(\gamma), \end{cases}$$

We conclude that K is Picard K-stable. □

3.5. Uniqueness of the Special Solution

Theorem 3. *We consider the Hilbert space $H = L^2((a,b) \times (0,k))$ that can be defined as the set of those functions:*

$$v : (a,b) \times [0,T] \to \mathbb{R}, \qquad \int\int uvdudv < \infty. \tag{40}$$

We now consider the following operator:

$$\eta(0,0),\eta=\begin{cases}(A_1-M_1-F_1)x_1(t)-L_1x_1^2(t)-\alpha\frac{x_1(t)x_2(t)}{1+x_1(t)}+G_1A_2x_2(t),\\(A_2-M_2-F_2)x_2(t)-L_2x_2^2(t)+\alpha\frac{x_1(t)x_2(t)}{1+x_1(t)}-G_1A_2x_2(t),\end{cases}$$

Proof. We prove that the inner product of:

$$(T(x_{11}(t)-x_{12}(t),x_{21}(t)-x_{22}(t),(\omega_1,\omega_2)), \tag{41}$$

where $(x_{11}(t)-x_{12}(t)),x_{21}(t)-x_{22}(t)$ are special solutions of the system. We can assume that $(x_{11}(t)-x_{12}(t))\cong x_{21}(t)-x_{22}(t)$.

Using the relationship between the norm and the inner function, we get:

$$\begin{aligned}&\left((A_1-M_1-F_1)(x_{11}(t)-x_{12}(t))-L_1(x_{11}(t)-x_{12}(t))^2-\alpha\frac{(x_{11}(t)-x_{12}(t))(x_{21}(t)-x_{22}(t))}{1+(x_{11}(t)-x_{12}(t))}+G_1A_2(x_{21}(t)-x_{22}(t)),\omega_1\right)\\&\leq(A_1-M_1-F_1)||x_{11}-x_{12}||||\omega_1||+L_1||x_{11}-x_{12}||^2||\omega_1||+\alpha\frac{||x_{11}(t)-x_{12}(t)||^2}{||1+(x_{11}(t)-x_{12}(t))||}||\omega_1||+G_1A_2||x_{11}-x_{12}||||\omega_1||\end{aligned} \tag{42}$$

and:

$$\begin{aligned}&\left((A_2-M_2-F_2)(x_{21}(t)-x_{22}(t))-L_2(x_{21}(t)-x_{22}(t))^2+\alpha\frac{(x_{11}(t)-x_{12}(t))(x_{21}(t)-x_{22}(t))}{1+(x_{11}(t)-x_{12}(t))}+G_1A_2(x_{22}(t)-x_{21}(t)),\omega_2\right)\\&\leq(A_2+M_2-F_2)||x_{21}-x_{22}||||\omega_2||+L_2||x_{22}-x_{21}||^2||\omega_2||+\alpha\frac{||x_{22}(t)-x_{21}(t)||^2}{||1+(x_{22}(t)-x_{21}(t))||}||\omega_2||+G_1A_2||x_{22}-x_{21}||||\omega_2||;\end{aligned} \tag{43}$$

for large number m and n, both solutions converge to the exact solution; if $\eta=||X_1-X_{11}||,||X_1-X_{12}||$ and $\nu=||x_2-x_{21}||,||x_2-x_{22}||$, we have:

$$\eta<\frac{\lambda_n}{2\left((A_1-M_1-F_1)+L_1||x_{11}(t)-x_{12}(t)||+\alpha\frac{||x_{11}(t)-x_{12}(t)||}{1+x_{11}(t)-x_{12}(t)}+G_1A_2\right)\left|\left|\omega_1\right|\right|}, \tag{44}$$

and:

$$\nu<\frac{\lambda_m}{2\left((A_2+M_2-F_2)+L_2||x_{22}(t)-x_{21}(t)||+\alpha\frac{||x_{22}(t)-x_{21}(t)||}{1-(x_{22}(t)-x_{21}(t))}+G_1A_2\right)\left|\left|\omega_2\right|\right|}, \tag{45}$$

where λ_n and λ_m are two very small positive parameters.

Using the topology concept, we conclude that $\lambda_n<0$ and $\lambda_m<0$, where:

$$\begin{aligned}&\left((A_1-M_1-F_1)+L_1||x_{11}(t)-x_{12}(t)||+\alpha\frac{||x_{11}(t)-x_{12}(t)||}{1+x_{11}(t)-x_{12}(t)}+G_1A_2\right)\neq 0,\\&\left((A_2+M_2-F_2)+L_2||x_{22}(t)-x_{21}(t)||+\alpha\frac{||x_{22}(t)-x_{21}(t)||}{1-(x_{22}(t)-x_{21}(t))}+G_1A_2\right)\neq 0.\end{aligned} \tag{46}$$

This completes the proof. □

Involving the Atangana–Baleanu fractional integral, we can propose a numerical solution using the Adams–Moulton rule:

$${}_0^{AB}\mathcal{I}_t^{\gamma}[f(t_{n+1})]=\frac{1-\gamma}{B(\gamma)}\left[\frac{f(t_{n+1})-f(t_n)}{2}\right]+\frac{\gamma}{\Gamma(\gamma)}\sum_{k=0}^{\infty}\left[\frac{f(t_{k+1})-f(t_k)}{2}\right]b_k^{\gamma}, \tag{47}$$

where:

$$b_k^{\gamma} = (k+1)^{1-\gamma} - (k)^{1-\gamma}. \tag{48}$$

Considering the above numerical scheme, we have:

$$\begin{aligned} x_{1(n+1)}(t) - x_{1(n)}(t) = x_{0(1)}^{n}(t) + \Bigg\{ \tfrac{1-\gamma}{B(\gamma)} \Bigg[(A_1 - M_1 - F_1) \tfrac{x_{1(n+1)}(t) - x_{1(n)}(t)}{2} \\ -L_1 \Big(\tfrac{x_{1(n+1)}^2(t) - x_{1(n)}^2(t)}{2} \Big) - \alpha \frac{\frac{(x_{1(n+1)}(t) - x_{1(n)}(t))(x_{2(n+1)}(t) - x_{2(n)}(t))}{2}}{1 + (x_{1(n+1)}(t) - x_{1(n)}(t))} + G_1 A_2 \Big(\tfrac{x_{2(n+1)} - x_{2(n)}(t)}{2} \Big) \Bigg] \\ + \tfrac{\gamma}{B(\gamma)} \textstyle\sum_{k=0}^{\infty} b_k^{\gamma} \cdot \Bigg[(A_1 - M_1 - F_1) \tfrac{x_{1(k+1)}(t) - x_{1(k)}(t)}{2} - L_1 \Big(\tfrac{x_{1(k+1)}^2(t) - x_{1(k)}^2(t)}{2} \Big) \\ -\alpha \frac{\frac{(x_{1(k+1)}(t) - x_{1(k)}(t))(x_{2(k+1)}(t) - x_{2(k)}(t))}{2}}{1 + (x_{1(k+1)}(t) - x_{1(k)}(t))} - G_1 A_2 \Big(\tfrac{x_{2(k+1)} - x_{2(k)}(t)}{2} \Big) \Bigg], \end{aligned} \tag{49}$$

and:

$$\begin{aligned} x_{2(n+1)}(t) - x_{2(n)}(t) = x_{0(2)}^{n}(t) + \Bigg\{ \tfrac{1-\gamma}{B(\gamma)} \Bigg[(A_2 - M_2 - F_2) \tfrac{x_{2(n+1)}(t) - x_{2(n)}(t)}{2} - \\ -L_2 \Big(\tfrac{x_{2(n+1)}^2(t) - x_{2(n)}^2(t)}{2} \Big) + \alpha \frac{\frac{(x_{1(n+1)}(t) - x_{1(n)}(t))(x_{2(n+1)}(t) - x_{2(n)}(t))}{2}}{1 + (x_{1(n+1)}(t) - x_{1(n)}(t))} + G_1 A_2 \Big(\tfrac{x_{2(n+1)} - x_{2(n)}(t)}{2} \Big) \Bigg] \\ + \tfrac{\gamma}{B(\gamma)} \textstyle\sum_{k=0}^{\infty} b_k^{\gamma} \cdot \Bigg[(A_2 - M_2 - F_2) \tfrac{x_{2(k+1)}(t) - x_{2(k)}(t)}{2} - L_2 \Big(\tfrac{x_{2(k+1)}^2(t) - x_{2(k)}^2(t)}{2} \Big) \\ -\alpha \frac{\frac{(x_{1(k+1)}(t) - x_{1(k)}(t))(x_{2(k+1)}(t) - x_{2(k)}(t))}{2}}{1 + (x_{1(k+1)}(t) - x_{1(k)}(t))} - G_1 A_2 \Big(\tfrac{x_{2(n+1)} - x_{2(n)}(t)}{2} \Big) \Bigg]. \end{aligned} \tag{50}$$

The proof of existence is described in detail by Alkahtani in [20].

4. Numerical Results

Example 1. *We present numerical simulations of the special solution of our model using the Adams–Moulton rule given by Equation* (23) *and Equations* (49) *and* (50) *for different arbitrary values of fractional order* γ. *We consider* $A_1 = 0.017$; $A_2 = 0.30$; $M_1 = 0.06$; $M_2 = 0.007$; $L_1 = 0.01$; $L_2 = 0.004$; $F_1 = 0.3$; $F_2 = 0.7$; $G_1 = 0.01$; $\alpha = 0.05$; *and initial conditions* $x(0) = 10$; $y(0) = 10$, *arbitrarily chosen. The simulation time is* 10 s, *and the step size used in evaluating the approximate solution was* $h = 0.001$. *The numerical results given in Figures 1a–d, 2a–d, 3a–d, and 4a–d show numerical simulations of the special solution of our model as a function of time for different values of* γ.

The figures show the interaction dynamics between a bilingual component and a monolingual component of a population in a particular environment. These numerical results show the influence of fractional order γ in the prediction between the unilingual and bilingual population.

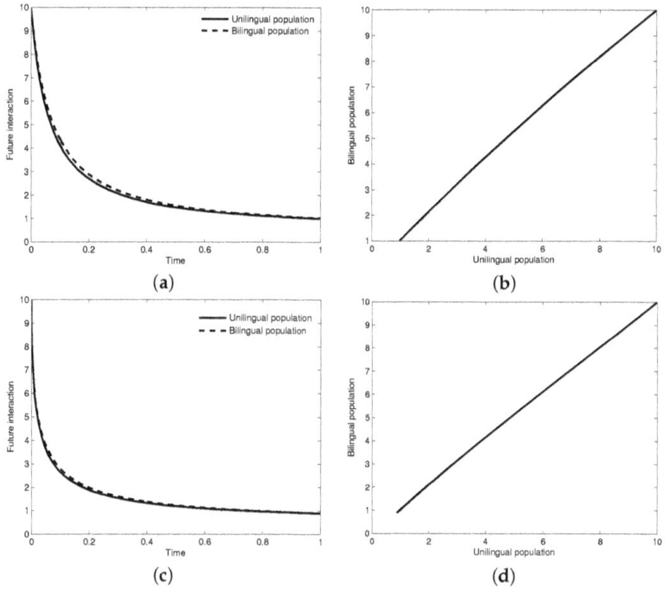

Figure 1. Numerical simulation for the nonlinear Freedman model via Liouville–Caputo fractional operator. In (**a**,**c**), the prediction between the two populations for $\gamma = 1$ (classical case) and $\gamma = 0.9$. In (**b**,**d**), the interaction between the unilingual and bilingual population for $\gamma = 1$ (classical case) and $\gamma = 0.9$.

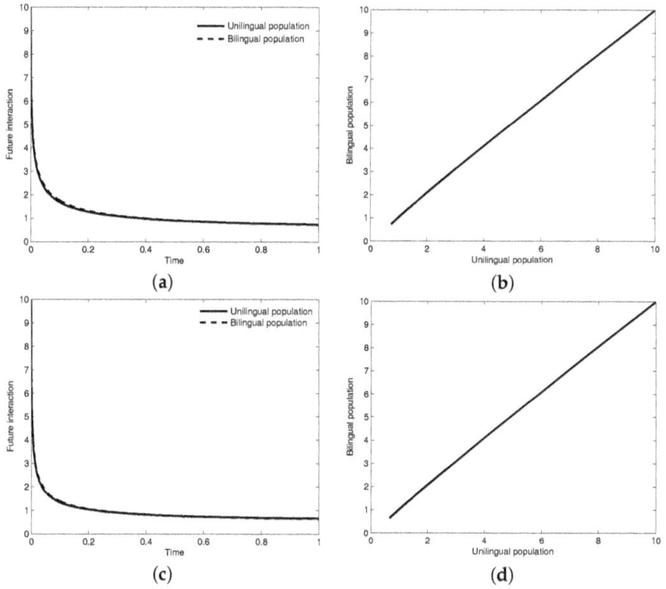

Figure 2. Numerical simulation for the nonlinear Freedman model via Liouville–Caputo fractional operator. In (**a**,**c**), the prediction between the two populations for $\gamma = 0.8$ and $\gamma = 0.7$. In (**b**,**d**), the interaction between the unilingual and bilingual population for $\gamma = 0.8$ (classical case) and $\gamma = 0.7$.

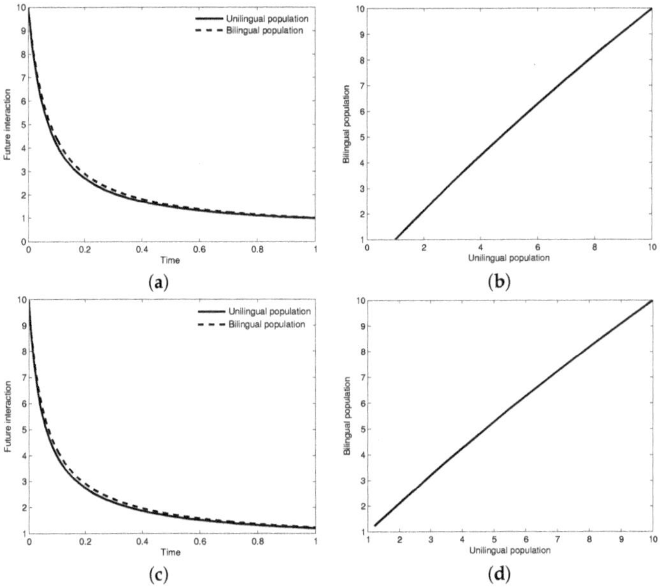

Figure 3. Numerical simulation for the nonlinear Freedman model via Atangana–Baleanu–Caputo fractional operator. In (**a**,**c**), the prediction between the two populations for $\gamma = 1$ (classical case) and $\gamma = 0.9$. In (**b**,**d**), the interaction between the unilingual and bilingual population for $\gamma = 1$ (classical case) and $\gamma = 0.9$.

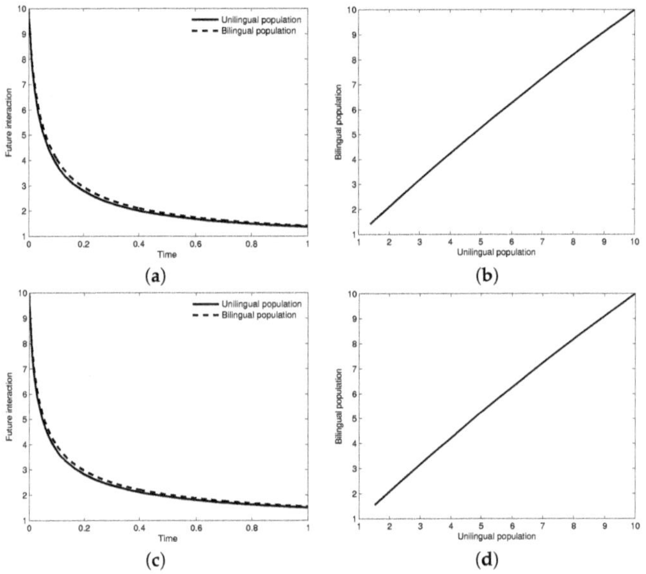

Figure 4. Numerical simulation for the nonlinear Freedman model via Atangana–Baleanu–Caputo fractional operator. In (**a**,**c**), the prediction between the two populations for $\gamma = 0.8$ and $\gamma = 0.7$. In (**b**,**d**), the interaction between the unilingual and bilingual population for $\gamma = 0.8$ (classical case) and $\gamma = 0.7$.

5. Conclusions

A Freedman model was considered using the fractional derivatives of the Liouville–Caputo and Atangana–Baleanu–Caputo types. The solutions of the alternative models were obtained using an iterative scheme based on the Laplace transform and the Sumudu transform. Furthermore, we employed the fixed point theorem to study the stability analysis of the iterative methods, and using properties of the inner product and the Hilbert space, the uniqueness of the special solution was presented in detail. Additionally, special solutions via the Adams–Moulton rule were obtained for both fractional derivatives. The results obtained using the Liouville–Caputo and Atangana–Baleanu-Caputo derivatives are exactly the same as the ordinary case. However, as γ takes values smaller than one, the results obtained become a little different, having a remarkable difference when $\gamma < 0.9$. This is due to the kernel involved in the definitions of the fractional derivative. The computer used for obtaining the results in this paper is an Intel Core i7, 2.6-GHz processor, 16.0 GB RAM (MATLAB R.2013a).

Acknowledgments: The authors appreciate the constructive remarks and suggestions of the anonymous referees that helped to improve the paper. We would like to thank to Mayra Martínez for the interesting discussions. José Francisco Gómez Aguilar acknowledges the support provided by CONACyT: cátedras CONACyT para jóvenes investigadores 2014. José Francisco Gómez Aguilar acknowledges the support provided by SNI-CONACyT.

Author Contributions: The analytical results were worked out by José Francisco Gómez Aguilar and Abdon Atangana. José Francisco Gómez Aguilar and Abdon Atangana wrote the paper. All authors have read and approved the final manuscript.

Conflicts of Interest: The authors declare no conflict of interest.

References

1. Baggs, I.; Freedman, H.I. A mathematical model for the dynamics of interactions between a unilingual and a bilingual population: Persistence versus extinction. *J. Math. Sociol.* **1990**, *16*, 51–75.
2. Baggs, I.; Freedman, H.I.; Aiello, W.G. Equilibrium characteristics in models of unilingual-bilingual population interactions. In *Ocean Wave Mechanics, Computational Fluid Dynamics, and Mathematical Modeling*; Rahman, M., Ed.; Computational Mechanics Publ.: Southampton, UK, 1990; pp. 879–886.
3. Duan, B.; Zheng, Z.; Cao, W. Spectral approximation methods and error estimates for Caputo fractional derivative with applications to initial-value problems. *J. Comput. Phys.* **2016**, *319*, 108–128.
4. Gómez, J.F. Comparison of the Fractional Response of a RLC Network and RC Circuit. *Prespacetime J.* **2012**, *3*, 736–742.
5. Ito, K.; Jin, B.; Takeuchi, T. On the sectorial property of the Caputo derivative operator. *Appl. Math. Lett.* **2015**, *47*, 43–46.
6. Kochubei, A.N. General fractional calculus, evolution equations, and renewal processes. *Integral Equ. Oper. Theory* **2011**, *71*, 583–600.
7. Sandev, T.; Metzler, R.; Tomovski, Z. Velocity and displacement correlation functions for fractional generalized Langevin equations. *Fract. Calc. Appl. Anal.* **2012**, *15*, 426–450.
8. Sandev, T.; Tomovski, Z.; Dubbeldam, J.L. Generalized Langevin equation with a three parameter Mittag–Leffler noise. *Phys. A Stat. Mech. Appl.* **2011**, *390*, 3627–3636.
9. Eab, C.H.; Lim, S.C. Fractional generalized Langevin equation approach to single-file diffusion. *Phys. A Stat. Mech. Appl.* **2010**, *389*, 2510–2521.
10. Ahmad, B.; Nieto, J.J.; Alsaedi, A.; El-Shahed, M. A study of nonlinear Langevin equation involving two fractional orders in different intervals. *Nonlinear Anal. Real World Appl.* **2012**, *13*, 599–606.
11. Metzler, R.; Klafter, J. Subdiffusive transport close to thermal equilibrium: from the Langevin equation to fractional diffusion. *Phys. Rev. E* **2000**, *61*, 6308–6311.
12. Lutz, E. Fractional langevin equation. In *Fractional Dynamics: Recent Advances*; World Scientific: Singapore, 2012; pp. 285–305.
13. Wiman, A. Über den Fundamentalsatz in der Teorie der Funktionen $E_a(x)$. *Acta Math.* **1905**, *29*, 191–201.
14. Prabhakar, T.R. A singular integral equation with a generalized Mittag–Leffler function in the kernel. *Yokohama Math. J.* **1971**, *19*, 7–15.

15. Shukla, A.K.; Prajapati, J.C. On a generalization of Mittag–Leffler function and its properties. *J. Math. Anal. Appl.* **2007**, 336, 797–811.
16. Srivastava, H.M.; Tomovski, Z. Fractional calculus with an integral operator containing a generalized Mittag–Leffler function in the kernel. *Appl. Math. Comput.* **2009**, *211*, 198–210.
17. Saxena, R.K.; Ram, J.; Vishnoi, M. Fractional differentiation and fractional integration of the generalized Mittag–Leffler function. *J. Indian Acad. Math.* **2010**, *32*, 153–162.
18. Kiryakova, V.S. Multiple (multiindex) Mittag–Leffler functions and relations to generalized fractional calculus. *J. Comput. Appl. Math.* **2000**, *118*, 241–259.
19. Atangana, A.; Baleanu, D. New Fractional Derivatives with Nonlocal and Non-Singular Kernel: Theory and Application to Heat Transfer Model. *Therm. Sci.* **2016**, *20*, 763–769.
20. Alkahtani, B.S.T. Chua's circuit model with Atangana–Baleanu derivative with fractional order. *Chaos Solitons Fractals* **2016**, *89*, 547–551.
21. Algahtani, O.J.J. Comparing the Atangana–Baleanu and Caputo-Fabrizio derivative with fractional order: Allen Cahn model. *Chaos Solitons Fractals* **2016**, *89*, 552–559.
22. Alkahtani, B.S.T. Analysis on non-homogeneous heat model with new trend of derivative with fractional order. *Chaos Solitons Fractals* **2016**, *89*, 566–571.
23. Gómez-Aguilar, J.F.; López-López, M.G.; Alvarado-Martínez, V.M.; Reyes-Reyes, J.; Adam-Medina, M. Modeling diffusive transport with a fractional derivative without singular kernel. *Phys. A Stat. Mech. Appl.* **2016**, *447*, 467–481.
24. Atangana, A.; Koca, I. Chaos in a simple nonlinear system with Atangana–Baleanu derivatives with fractional order. *Chaos Solitons Fractals* **2016**, *89*, 447–454.
25. Atangana, A.; Koca, I. On the new fractional derivative and application to nonlinear Baggs and Freedman model. *J. Nonlinear Sci. Appl.* **2016**, *9*, 2467–2480.
26. Wyburn, J.; Hayward, J. The future of bilingualism: an application of the Baggs and Freedman model. *J. Math. Sociol. 2008*, *32*, 267–284.
27. Caputo, M.; Mainardi, F. A new dissipation model based on memory mechanism. *Pure Appl. Geophys.* **1971**, *91*, 134–147.
28. Watugala, G.K. Sumudu transform: A new integral transform to solve differential equations and control engineering problems. *Integr. Educ.* **1993**, *24*, 35–43.
29. Li, C.; Tao, C. On the fractional Adams method. *Comput. Math. Appl.* **2009**, *58*, 1573–1588.
30. Diethelm, K.; Ford, N.J.; Freed, A.D. Detailed error analysis for a fractional Adams method. *Numer. Algorithms* **2004**, 36, 31–52.
31. Katatbeh, Q.K.; Belgacem, F.B.M. Applications of the Sumudu transform to fractional differential equations. *Nonlinear Stud.* **2011**, *18*, 99–112.
32. Bulut, H.; Baskonus, H.M.; Belgacem, F.B.M. The analytical solutions of some fractional ordinary differential equations by Sumudu transform method. *Abstr. Appl. Anal.* **2013**, *2013*, 203875.

© 2018 by the authors. Licensee MDPI, Basel, Switzerland. This article is an open access article distributed under the terms and conditions of the Creative Commons Attribution (CC BY) license (http://creativecommons.org/licenses/by/4.0/).

fractal and fractional

MDPI

Article

Comparison between the Second and Third Generations of the CRONE Controller: Application to a Thermal Diffusive Interface Medium

Xavier Moreau [1], Roy Abi Zeid Daou [2,3,*] and Fady Christophy [4]

1 UMR 5218 CNRS, IMS Laboratory, University of Bordeaux, 33405 Talence Cedex, France; xavier.moreau@u-bordeaux.fr
2 Biomedical Technologies Department, Faculty of Public Health, Lebanese German University, Sahel Alma, Jounieh 1200, Lebanon
3 MART Learning, Education and Research Center, Chananiir 1200, Lebanon
4 Faculty of Engineering, Lebanese University, Tripoli 1300, Lebanon; fadychris@hotmail.com
* Correspondence: r.abizeiddaou@lgu.edu.lb or roydaou@mart-ler.org; Tel.: +961-3-396-099

Received: 19 December 2017; Accepted: 12 January 2018; Published: 17 January 2018

Abstract: The control of thermal interfaces has gained importance in recent years because of the high cost of heating and cooling materials in many applications. Thus, the main focus in this work is to compare the second and third generations of the CRONE controller (French acronym of Commande Robuste d'Ordre Non Entier), which means a non-integer order robust controller, and to synthesize a robust controller that can fit several types of systems. For this study, the plant consists of a rectangular homogeneous bar of length L, where the heating element in applied on one boundary, and a temperature sensor is placed at distance x from that boundary (x is considered very small with respect to L). The type of material used is the third parameter, which may help in analyzing the robustness of the synthesized controller. The originality of this work resides in controlling a non-integer plant using a fractional order controller, as, so far, almost all of the systems where the CRONE controller has been implemented were of integer order. Three case studies were defined in order to show how and where each CRONE generation controller can be applied. These case studies were chosen in such a way as to influence the asymptotic behavior of the open-loop transfer function in the Black–Nichols diagram in order to point out the importance of respecting the conditions of the applications of the CRONE generations. Results show that the second generation performs well when the parametric uncertainties do not affect the phase of the plant, whereas the third generation is the most robust, even when both the phase and the gain variations are encountered. However, it also has some limitations, especially when the temperature to be controlled is far from the interface when the density of flux is applied.

Keywords: CRONE controller; homogeneous plan diffusive interface; semi-infinite medium; robust control

1. Introduction

The thermal application is a common application for control. It is used in several engineering domains as the heating/cooling of houses [1,2], the control of the temperature inside the car [3,4] or in the industrial machinery [5], and much more. In more details, the work in this field started in the 1940s with the works of Jones [6] and a group of American electrical engineers [7]. Then, in the 1950s, the researchers were more involved in the control of thermal neutron reactors [8,9] and electric cables dissipation [10,11]. These studies had increased since the end of the previous decade, when the scientists stopped focusing on the study of the material properties, and started searching for more ecological systems to control the temperature within the building in order to reduce the oil and fuel usage [12,13].

However, the work on diffusive interfaces was launched in the late 1970s and 1980s with the works of Kumar, who modeled the thermal boundaries near an oil plant [14]. As for the others, they worked on the modeling of diffusive interfaces at furnaces [15], boilers [16] and some geometric shapes representing the separation medium [17,18]. In the last decade of the twentieth century, this domain was investigated by lots of researchers who demonstrated that the relation between the input flux and the output temperature is of a fractional order. From among them, we can recognize the remarkable works of Trigeassou and his team [19–21], Battablia [22,23], the CRONE team of Bordeaux university [24–26], the relevant modeling approach based on the space-fractional continuum models [27–29], and much more.

As for the control, it was applied a long time ago, even before Christ, when Ktesibios (270 B.C.) implemented the water clocks working on feedback control [30]. The contemporary control theory was effectively launched in the middle of the twentieth century with the remarkable work of Bode, Nichols, and Nyquist [31]. In its first years, the integer and linear order systems were treated on both time and frequency domains.

As for fractional theory, it is relatively an old idea that dates back to the end of the seventieth century, when some letters were exchanged between two well-known mathematicians at that epoch, L'Hopital and Leibnitz in which they asked each other about the meaning of a derivative of order 0.5 [32]. Many mathematicians defined this type of derivative/integration as Liouville, Caputo, and Riemann [33]. Nevertheless, these calculations remained theoretical until the last quarter of the twentieth century, when Oustaloup was the first to design a fractional control of order 3/2 to control a laser beam [34]. He was one of the first researchers to introduce the fractional calculus in engineering domain applications [35–38]. Due to the obtained results, many other contributions appeared from Ortigueira [39,40], Machado [41,42], and Vinagre [43,44].

Hence, this paper will present the control of a diffusive interface medium, consisting of a homogeneous rectangular finite rod, by the CRONE controller. The novelty of this work is that it presents a fractional order plant that will be controlled by a fractional order controller, who will be, for the first time, real (when applying the second generation CRONE), and in a second time, this controller will be complex (by applying the third generation CRONE). Added to that, the robustness of these controllers will be studied when varying the bar metal, the position of the point where the temperature is to be measured, the length of the bar, and so on. Thus, three case studies will be shown in order to show the behavior of the open-loop system, and to indicate the conditions of application for each CRONE generation. Since the objective of this paper is to point out the importance of applying the CRONE controller to a fractional order plant, we were not interested in comparing the performance of this controller to other ones.

For the remaining part of this paper, we will consider that the length of the bar is much more important than the position of the temperature sensor. As for the user specifications, two main constraints will be considered hereafter:

- A crossover frequency $\omega_{cg} = 1$ rad/s;
- A phase margin $M_{\phi} = 3$ dB;

To do so, this paper will be divided as follows: in Section 2, the homogeneous finite and rectangular bar will be presented, along with its mathematical modeling and its boundary conditions. In Section 3, the different applied controllers will be introduced; an overview of the three CRONE generations will also be proposed, along with the conditions of application of each generation. In Section 4, the second CRONE generation will be presented along with two case studies: the first one shows an ideal domain of application where the phase is constant and the gain varies whenever plant parameters change. The second case presents both phase and gain variations, and the use of the already synthesized controller appears to be no more robust. Hence, for this second case study, the third generation controller will be introduced in Section 5, and a performance analysis will be presented. At the end, Section 6 will provide a conclusion and some future works to enrich this work.

2. Plant Modeling

The homogeneous diffusive interface medium will be presented in this part. The test bench is constituted of a rectangular homogeneous rod with a square section S. It may consist of aluminum, copper, or iron, and it is of a finite distance L, as shown in Figure 1. The characteristics of the medium, its differential equations, and the boundary conditions, along with the features of the material used will be proposed hereafter.

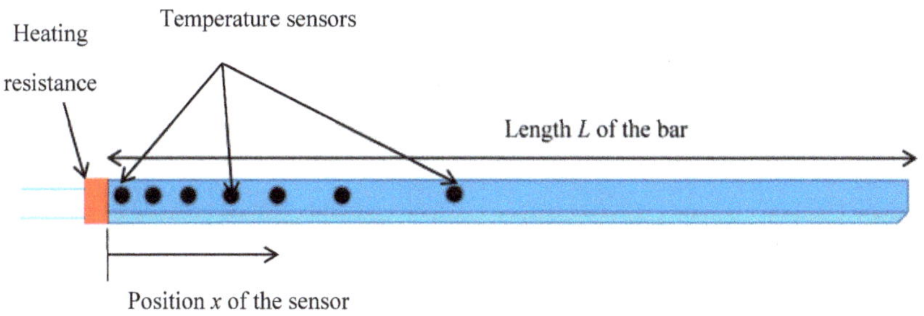

Figure 1. Representation of diffusive interface medium along with the sensors and the heating element.

2.1. Partial Differential Equations (PDE)

The input of this medium will be the density of flux $\varphi(t)$ (which is equal to the flux $\phi(t)$ divided by the section S of the source), whereas the output is the temperature at a location x. The partial differential equations of this medium are shown in system (1), where the first equation shows the temperature value for any value of x. As for the second equation, it represents the temperature variation at the boundary where the heating element is applied.

$$\begin{cases} \frac{\partial T(x,t)}{\partial t} = \alpha_d \frac{\partial^2 T(x,t)}{\partial x^2}, \; x > 0, \; t > 0 \\ -\lambda \frac{\partial T(x,t)}{\partial x} = \phi(t), \; x = 0, \; t > 0 \end{cases}, \tag{1}$$

where α_d represents the thermal diffusivity of the material, and $T(x,t)$ represents the temperature at the position x for time t.

In the second boundary equation, the flux is not applied, and the temperature along the bar at the initial time (t = 0 s) is expressed by the following system:

$$\begin{cases} -\lambda \frac{\partial T(x,t)}{\partial x} = 0, \; x = L, \; t > 0 \\ T(x,t) = 0, \; 0 \leq x < L, \; t = 0 \end{cases}. \tag{2}$$

2.2. Plant Transfer Function

Based on the partial differential equations that represent the relation between the applied density of flux $\varphi(t)$ and the temperature variation $T(x,t,L)$ within the diffusive interface medium of length L, one can deduce the transfer function $H(x,s,L)$ that models this system (interested authors can refer to the following references for a detailed calculation of this transfer function [45,46]). Note here that the temperature varies with respect to the material used, the placement x of the temperature sensor, and the length L of the rod. This transfer function $H(x,s,L)$ is given by:

$$
\begin{aligned}
H(x,s,L) &= \frac{\overline{T}(x,s,L)}{\overline{\varphi}(s)} = H_0 \frac{1}{s^{0.5}} \frac{1}{\tanh\left(\sqrt{\frac{s}{\omega_L}}\right)} \frac{\cosh\left(\sqrt{\frac{s}{\omega_{Lx}}}\right)}{\cosh\left(\sqrt{\frac{s}{\omega_L}}\right)}, \\
H_0 &= \frac{1}{S\,\eta_d},\ \eta_d = \sqrt{\lambda\,\rho\,C_p},\ \omega_L = \frac{\alpha_d}{L^2},\ \omega_{Lx} = \frac{\alpha_d}{(L-x)^2},
\end{aligned}
\quad (3)
$$

where $\overline{T}(x,s,L)$ and $\overline{\varphi}(s)$ represent the Laplace transform of $T(x,t,L)$ and $\varphi(t)$ respectively, η_d is the thermal effusivity, λ represents the thermal conductivity, ρ is the medium density, and C_p is the medium heat.

From Equation (3), one can notice the fractional order operator ($\sqrt{s}$) residing in the hyperbolical trigonometric functions, as well as the pure semi-integrator. This can confirm, once again, the novelty of this work by applying a fractional order control (the CRONE controller) to, most importantly, a non-integer order plant.

However, the expression of the transfer function $H(x,s,L)$ can be reduced: in fact, as the position of the temperature sensor is neglected with respect to the length of the bar ($x << L$), one can deduce that $\omega_{Lx} \approx \omega_L$. Thus, Equation (3) can be written as follows, where this latter defines the validation model of the system:

$$
H(x,s,L) = \frac{\overline{T}(x,s,L)}{\overline{\varphi}(s)} = H_0 \frac{1}{s^{0.5}} \frac{1}{\tanh\left(\sqrt{\frac{s}{\omega_L}}\right)} e^{-\sqrt{\frac{s}{\omega_x}}}, \quad (4)
$$

where $\omega_x = \alpha_d/x^2$.

2.3. Material Characteristics

As previously proposed, three materials will be used: copper (Cop), iron (Iro), and aluminum (Alu). The last one will be considered for the nominal case. Table 1 shows the characteristics of the three materials, along with the values of the two transitional frequencies ω_L and ω_x for three values of the length, L, and the temperature sensor position, x.

Table 1. Physical characteristics of the used materials. Alu: aluminum; Cop: copper; Iro: iron.

Material	α_d	η_d	H_0	ω_L (rad/s)			ω_x (rad/s)		
	m^2/s	$W{\cdot}K^{-1}{\cdot}m^{-2}{\cdot}s^{0.5}$	$K{\cdot}s^{0.5}{\cdot}W^{-1}$	$L = 0.25$ m	$L = 0.5$ m	$L = 1$ m	$x = 0$ cm	$x = 0.5$ cm	$x = 1$ cm
Cop.	117×10^{-6}	3.72×10^4	0.269	19×10^{-4}	4.68×10^{-4}	1.17×10^{-4}	Infinite	4.68	1.17
Alu.	97×10^{-6}	2.41×10^4	0.416	16×10^{-4}	3.88×10^{-4}	0.97×10^{-4}	Infinite	3.88	0.97
Iro.	23×10^{-6}	1.67×10^4	0.596	3.68×10^{-4}	0.92×10^{-4}	0.23×10^{-4}	Infinite	0.92	0.23

When synthesizing the controllers, two case studies will be proposed based on the values of L and x for the three material types. In the next sections, these case studies will be defined in more detail, and the plant uncertainties will be modeled.

3. CRONE Controllers Presentation

The controllers that would be used in this application are the second and third generation CRONE controllers. In the following, we will present each controller, and the method to synthesize it. The two next sections will show the applications of the second and the third generations using two different case studies. This will help the user understand the conditions in which to apply each of the CRONE generations.

The CRONE controller is the first fractional order controller developed. It was launched in 1975, and it was introduced using three generations. We will focus in this paper on the second and third generations, knowing that the first two generations were treated in previous works [47,48]. However, a brief overview over the three generations will be presented below.

The first generation Crone controller proposes to use a controller without phase variation around crossover frequency ω_{cg}. Thus, the phase margin variation only results from the plant variation. This strategy has to be used when frequency ω_{cg} is within a frequency range where the plant phase is constant. In this range, the plant variations are only gain-like. This first generation uses the a priori calculation where the controller transfer function is calculated directly based on the user specifications.

The second generation CRONE control is applied when the plant variations are gain-like around the gain crossover frequency ω_{cg}, and the plant phase variation is canceled by those of the controller. Thus, there is no phase margin variation when the frequency of ω_{cg} varies. Such a controller produces a constant open-loop phase whose Nichols locus is a vertical straight line named the frequency template. This controller is synthesized a posteriori, where its transfer function is deduced from the open-loop transfer function.

The third generation Crone controller is used when the plant frequency uncertainty domains are of various types (not only gain-like, but present both gain and phase variation). The vertical template is then replaced by a generalized template or by a multi-template (or curvilinear template) defined by a set of generalized templates. Here also, the transfer function is defined a posteriori based on the open-loop behavior [49].

4. Second CRONE Generation

4.1. Introduction

As the transfer function of this generation is synthesized a posteriori, the transfer function of the open-loop system is defined as follows (for frequencies in the range of $[\omega_A, \omega_B]$):

$$\beta(s) = \left(\frac{\omega_{cg}}{s}\right)^n, \tag{5}$$

where ω_{cg} is the frequency for which the uncertainties do not lead to any phase variation, $n \in \mathbb{R}$ and $n \in [1,2]$.

The complementary sensitivity function $T(s)$ and the sensitivity function $S(s)$ are defined as follows:

$$T(s) = \frac{\beta(s)}{1+\beta(s)} = \frac{1}{1+\left(\frac{s}{\omega_{cg}}\right)^n} \text{ and } S(s) = \frac{1}{1+\beta(s)} = \frac{\left(\frac{s}{\omega_{cg}}\right)^n}{1+\left(\frac{s}{\omega_{cg}}\right)^n}. \tag{6}$$

Around the crossover frequency ω_{cg}, the Black–Nichols plot of the open-loop transfer function $\beta(s)$ is a vertical asymptote with a constant phase equal to n, as shown in Figure 2. This asymptote allows having [48]:

- a robust phase margin M_ϕ equal to $(2-n)\times\pi/2$;
- a robust resonance factor Q_T, defined as follows:

$$Q_T = \frac{\sup_\omega |T(j\omega)|}{|T(j0)|} = \frac{1}{\sin(n\pi/2)}; \tag{7}$$

- a robust gain module M_m, defined as follows:

$$M_m = \inf_\omega |\beta(j\omega)+1| = \left(\sup_\omega |S(j\omega)|\right)^{-1} = \sin(n\pi/2); \tag{8}$$

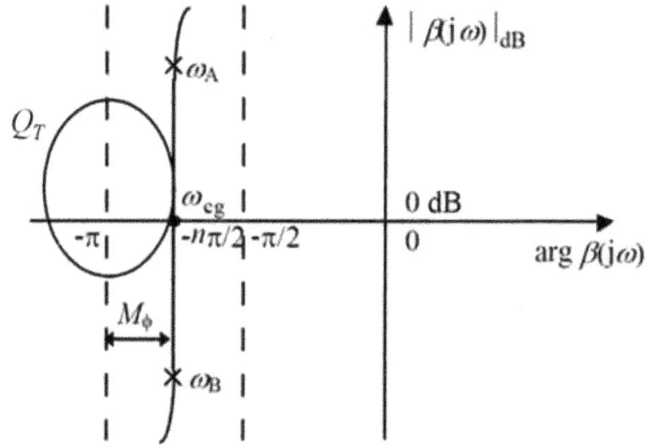

Figure 2. Asymptotic behavior of the closed loop transfer function in the Black–Nichols diagram.

As for the control signal and the transient error, Equation (5) can be truncated in frequency, and a low pass filter as well as an integrator must be added. Hence, the new form of the open-loop transfer will appear as follows:

$$\beta(s) = \beta_0 \left(\frac{1+s/\omega_l}{s/\omega_l}\right)^{n_l} \left(\frac{1+s/\omega_h}{1+s/\omega_l}\right)^{n} (1+s/\omega_h)^{-n_h}, \tag{9}$$

where ω_l and ω_h represent the low and high transitional frequencies, n is the fractional order (varying between 1 and 2) around the frequency ω_{cg}, n_l and n_h are the system behavior at low and high frequencies, and β_0 is a constant value that assures a crossover frequency ω_{cg}. It is expressed in Equation (10):

$$\beta_0 = (\omega_{cg}/\omega_l)^{n_l} \left(1+(\omega_{cg}/\omega_l)^2\right)^{(n-n_l)/2} \left(1+(\omega_{cg}/\omega_h)^2\right)^{(n_h-n)/2}. \tag{10}$$

Figure 3 shows the asymptotic behavior in a Bode diagram for this open-loop transfer function $\beta(s)$. The fractional order behavior is defined over the interval $[\omega_A, \omega_B]$, and it belongs to the nominal crossover frequency ω_{cgnom}. In order to respect the robustness of the stability degree, it is necessary to define the margins for ω_{cg}, such as:

$$\forall\, \omega_{cg} \in \left[\omega_{cg\min}; \omega_{cg\max}\right], \omega_A \le \omega_{cg} \le \omega_B \quad \Rightarrow \quad \begin{cases} \omega_A \le \omega_{cg\min} \\ \omega_B \ge \omega_{cg\max} \end{cases}. \tag{11}$$

As shown in Figure 3, two new cutoff frequencies are introduced (ω_l and ω_h), which help getting the fractional order behavior between ω_A and ω_B. Previous studies have shown that it is sufficient for ω_l to be one decade less ω_B, whereas for ω_l, it must be one decade above ω_h [49].

$$\begin{cases} \omega_l = \omega_A/10 \\ \omega_h = 10\,\omega_B \end{cases}. \tag{12}$$

Thus, one can define ω_{cgnom} as being the geometric median of ω_l and ω_h. Added to that, a new parameter r, being the ratio of ω_B and ω_A, is introduced:

$$\begin{cases} \sqrt{\omega_l\,\omega_h} = \omega_{cgnom} \\ r = \frac{\omega_B}{\omega_A} \end{cases}. \tag{13}$$

As a consequence, ω_l and ω_h could be written with respect to ω_{cgnom} and r as follows:

$$\begin{cases} \sqrt{\omega_l\,\omega_h} = \omega_{cgnom} \\ \frac{\omega_h}{\omega_l} \;=\; 100\,r \end{cases} \quad \Rightarrow \quad \begin{cases} \omega_l \;=\; \omega_{cgnom}/\left(10\,\sqrt{r}\right) \\ \omega_h \;=\; \omega_{cgnom}\,10\,\sqrt{r} \end{cases}. \tag{14}$$

Once the open-loop transfer function is calculated, one can conclude the CRONE controller transfer function as being the ratio of the open-loop transfer function $\beta(j\omega)$ over the nominal plant's transfer function $P_0(j\omega)$:

$$C_F(j\omega) = \beta(j\omega)/P_0(j\omega). \tag{15}$$

A last step is always needed in order to pass from the fractional form $C_F(j\omega)$ to the rational form $C_R(j\omega)$. Several methods could be applied in this case; however, one simple method is based on the representation of the function using a recursive distribution of poles and zeros. Each pole and zero form a cell. The higher the number of cells, the most accurate the results are, but the more complex the transfer function would be. However, four to eight cells would be enough, as the fractional frequency range is below three decades [50]. Another option exists in using the CRONE toolbox, which can give the rational representation of the fractional form since it knows the frequency response of $C_F(j\omega)$ [51,52].

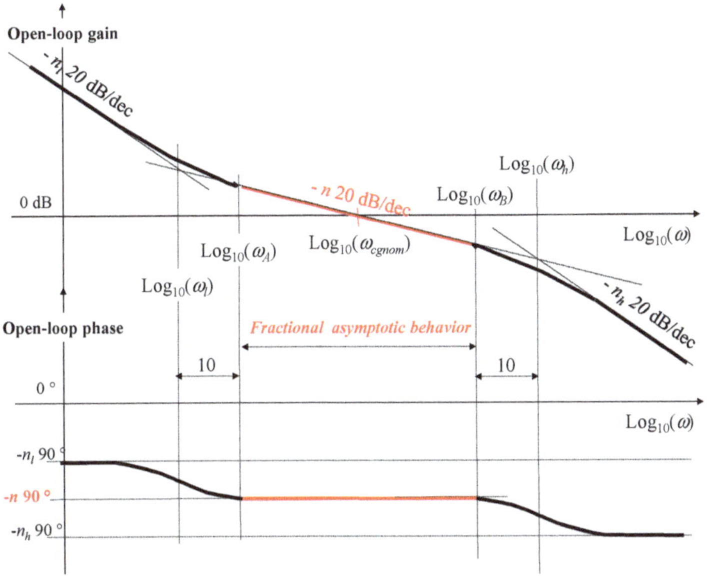

Figure 3. Asymptotic behavior of the open-loop transfer function $\beta(s)$ in the Bode diagram.

4.2. *First Case Study*

4.2.1. Plant Parameters

As the objectives of the second generation of the CRONE controller is to have a constant phase with a variable gain when varying the parameters of the plant, the choice of the values of L and x is crucial in order to apply this generation. Hence, these values are listed below:

- Aluminum, L = 1 m and x = 0.5 cm $\rightarrow \omega_L = 0.97\,10^{-4}$ rad/s and ω_x = 3.88 rad/s;
- Copper, L = 1.1 m and x = 0.55 cm $\rightarrow \omega_L = 0.97\,10^{-4}$ rad/s and ω_x = 3.87 rad/s;
- Iron, L = 0.49 m and x = 0.243 cm $\rightarrow \omega_L = 0.96\,10^{-4}$ rad/s and ω_x = 3.89 rad/s.

Figure 4 shows the Bode diagrams for the three plants. The phase constancy for the three outputs is well noted; however, the changes in the gain are also observed. Hereafter, the aluminum will be considered the nominal case, whereas the copper and iron will be considered the extreme cases.

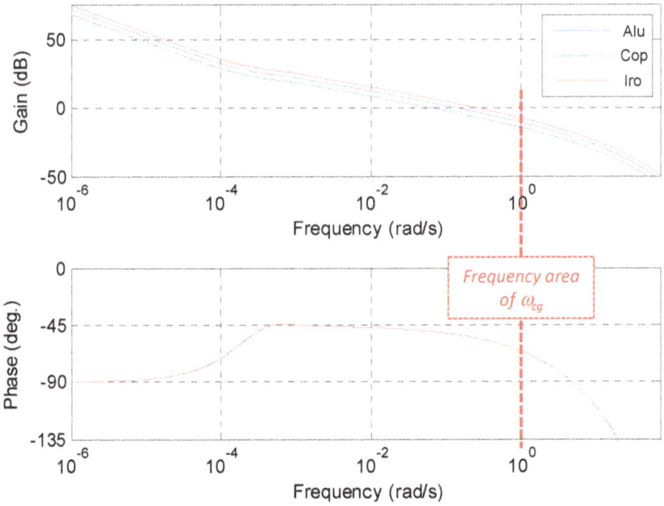

Figure 4. Bode plots of $H(x,j\omega,L)$ for aluminum (in blue), copper (in green), and iron (in red) for the first case study.

4.2.2. Synthesis Model

The plant transfer function $H(x,s,L)$, which was already presented in Equations (3) and (4), could be written in another way in order to facilitate the computation of the controller transfer function. Thus, it could be expressed as follows based on the approximation for the tanh function. The new model $P_2(s)$ in Equation (16) will be considered as the synthesis model that will be easier to use in order to calculate the controller transfer function. Interested readers can refer to a previous work of the authors for more details about the calculations [53].

$$P_2(s) = H_0^* \frac{(1+s/\omega_L)^{0.5}}{s/\omega_L} e^{-\sqrt{\frac{s}{\omega x}}}, \tag{16}$$

where

$$H_0^* = \frac{H_0}{\omega_L^{0.5}}. \tag{17}$$

Figure 5 represents the Bode plots of $P_2(j\omega)$ (in blue) and of $H(x,j\omega,L)$ (in green) obtained with aluminum for $L = 1$ m and $x = 0.5$ cm. It is well noted the coherence of both plots (for the gain and the phase) in the frequency range around ω_{cg}.

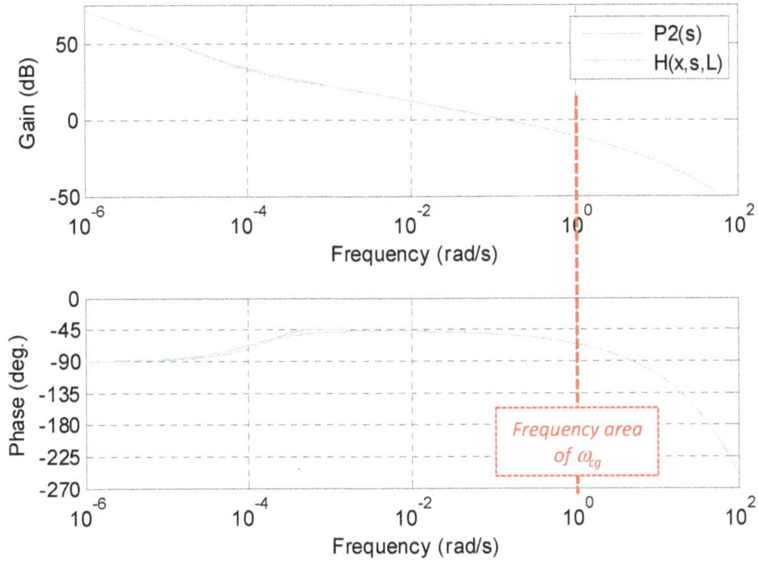

Figure 5. Bode plots of $P_2(j\omega)$ (in blue) and of $H(x,j\omega,L)$ (in green) obtained with aluminum for L = 1 m and x = 0.5 cm.

4.2.3. Controller Transfer Function

As already discussed, the synthesis of the CRONE controller transfer function is done a posteriori. Thus, we have first to compute the open-loop transfer function (Equation (9)), then by replacing the plant's transfer function by its value, we will obtain the controller transfer function, which could be expressed as follows:

$$C_F(s) = \beta_0 \left(\frac{1+s/\omega_l}{s/\omega_l}\right)^{n_l} \left(\frac{1+s/\omega_h}{1+s/\omega_l}\right)^{n} \frac{1}{\left(1+s/\omega_h\right)^{n_h}} \frac{s/\omega_L}{H_0^*\left(1+s/\omega_L\right)^{0.5}} e^{\sqrt{\frac{s}{\omega x}}}. \tag{18}$$

Taking into consideration the specifications of the user guide, the different variables can be defined. Thus,

- n_l = 2, in order to assure a null training error;
- n_h = 1.5, in order to limit the input sensitivity;
- Q_T = 3 dB or $M_\phi = 45° \rightarrow n = (180° - M_\phi)/90° = 1.5$;
- $\omega_{cgnom} = 1\ \text{rad/s}$;

So, taking into account the previous values, Equation (18) can be written as follows:

$$C_F(s) = C_0\left(\frac{\omega_l}{s}\right) \frac{\left(1+s/\omega_l\right)^{2-n}}{\left(1+s/\omega_h\right)^{1.5-n}\left(1+s/\omega_L\right)^{0.5}} e^{\sqrt{\frac{s}{\omega x}}}, \tag{19}$$

where:

$$C_0 = \frac{\beta_0\,\omega_l}{H_0^*\,\omega_L}. \tag{20}$$

In order to get rid of the exponent, a Taylor development was needed for e^z when z tends towards zero.

$$\lim_{z \to 0} e^z = 1 + z + \frac{z^2}{2!} + \frac{z^3}{3!} + \ldots + \frac{z^k}{k!} + \ldots = \sum_{k=0}^{\infty} \frac{z^k}{k!}, \tag{21}$$

When truncated at order 2 and considering that $z = (s/\omega_x)^{0.5}$, one can obtain:

$$e^{\sqrt{\frac{s}{\omega_x}}} \approx 1 + \left(\frac{s}{\omega_x}\right)^{0.5} + \frac{s}{2\,\omega_x}. \tag{22}$$

So, the fractional approximated form $\widetilde{C}_F(s)$ is defined by:

$$\begin{aligned}\widetilde{C}_F(s) &= C_0 \left(\frac{\omega_l}{s}\right) \frac{(1 + s/\omega_l)^{2-n}}{(1 + s/\omega_h)^{1.5-n} (1 + s/\omega_L)^{0.5}} \left(1 + \left(\frac{s}{\omega_x}\right)^{0.5} + \frac{s}{2\,\omega_x}\right) \\ &= \frac{C_0^*}{s} \left(1 + \left(\frac{s}{\omega_x}\right)^{0.5} + \frac{s}{2\,\omega_x}\right)\end{aligned} \tag{23}$$

where $C_0^* = C_0\,\omega_l = 2.405\,\mathrm{W \cdot s^{-1} \cdot deg^{-1}}$, and $\omega_x = 3.88$ rad/s.

Referring to Equation (23), the controller is constituted of a simple integrator, which allows removing the noise caused by the exponent around the crossover frequency ω_{cg}. For the low frequencies, the controller has an integrator behavior, whereas for high frequencies, it has a proportional behavior.

In order to apply this controller, the rational form is needed. For this example, we will be using the cascade representation, as shown in Equation (24). The use of the CRONE toolbox helps defining the values of the poles and the zeros. The new approximated transfer function will be as follows:

$$C_R(s) = \frac{C_0^*}{s} \frac{\prod_{j=1}^{3} \left(1 + \frac{s}{\omega_{zj}}\right)}{\prod_{j=1}^{2} \left(1 + \frac{s}{\omega_{pj}}\right)}, \tag{24}$$

where:

$$\begin{cases} \omega_{z1} = 0.8\,\mathrm{rad/s} & \omega_{p1} = 1.25\,\mathrm{rad/s} \\ \omega_{z2} = 6.5\,\mathrm{rad/s} & \omega_{p2} = 29\,\mathrm{rad/s} \\ \omega_{z3} = 55\,\mathrm{rad/s} & \end{cases}. \tag{25}$$

4.2.4. Performance Analysis

Concerning the behavior of the second generation CRONE controller applied to a plant where the uncertainties are modeled only by gain variation and an invariable but non-constant phase for the three plants, the following plots will resume the outcome.

Figure 6 represents the Bode diagrams for the different controllers transfer functions: the fractional form $\widetilde{C}_F(s)$ (in blue), and the rational form $C_R(s)$ (in green).

Figure 7 shows the open-loop Black–Nichols plots (a), the closed-loop step responses (b), and their control inputs for a step input of 1 °C (c) for aluminum (in blue), copper (in green), and iron (in red) for the $C_R(s)$ controller.

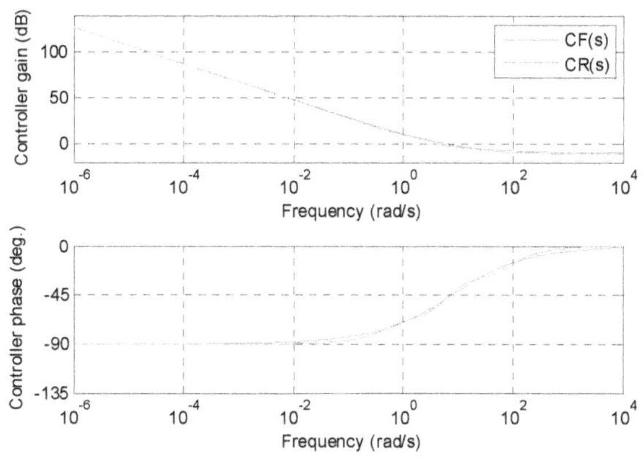

Figure 6. Bode diagrams for the fractional controller $\widetilde{C}_F(s)$ (in blue) and the rational form $C_R(s)$ (in green).

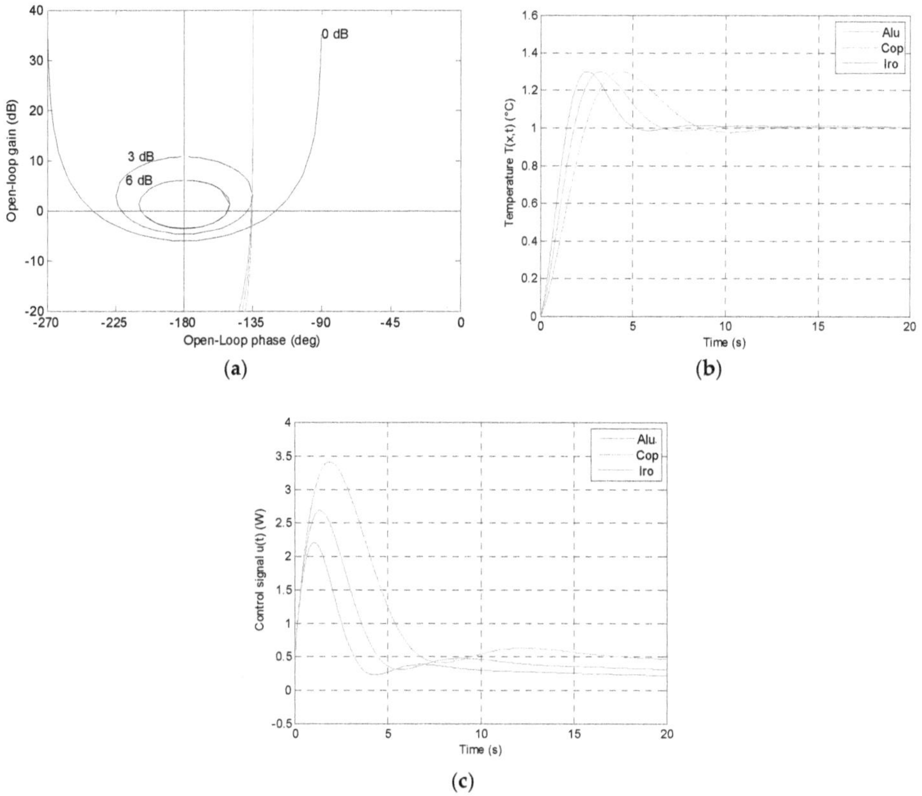

Figure 7. Open-loop Black–Nichols plots (**a**), closed-loop step responses (**b**), control inputs for a step input of 1 °C (**c**) for aluminum (in blue), copper (in green), and iron (in red).

As a conclusion, the exact coherence between the fractional and the rational transfer functions is well noted (Figure 6). Concerning the robustness of the controller, one can see clearly that the three open-loop transfer functions in the Nichols diagram are tangent to the same contour (3 dB). As for the closed-loop step responses, all three outputs have the same first overshoot value and the same damping ratio. Thus, we can confirm that the second generation CRONE controller is robust whenever the conditions of application of this generation are met.

4.3. Second Case Study

In this second case study, we will choose L and x arbitrarily in such a way that the plant will present both phase and gain variations when changing the parameters of the plant, and we will apply the previously synthesized controller (Equations (24) and (25) to study the performance of the new system.

4.3.1. Plant Parameters

The values of L and x chosen for this case study are as follows:

- Aluminum, L = 1 m and x = 0.5 cm $\rightarrow \omega_L = 0.97\ 10^{-4}$ rad/s and $\omega_x = 3.88$ rad/s;
- Copper, L = 1.1 m and x = 1 cm $\rightarrow \omega_L = 0.97\ 10^{-4}$ rad/s and $\omega_x = 1.17$ rad/s;
- Iron, L = 0.49 m and x = 0.1 cm $\rightarrow \omega_L = 0.96\ 10^{-4}$ rad/s and $\omega_x = 23$ rad/s.

Figure 8 shows the Bode diagrams for the three plants. Both the phase and the gain vary when changing the parameters of the plant. As for the first case study, aluminum will be considered as the nominal case, whereas copper and iron will be considered the extreme cases.

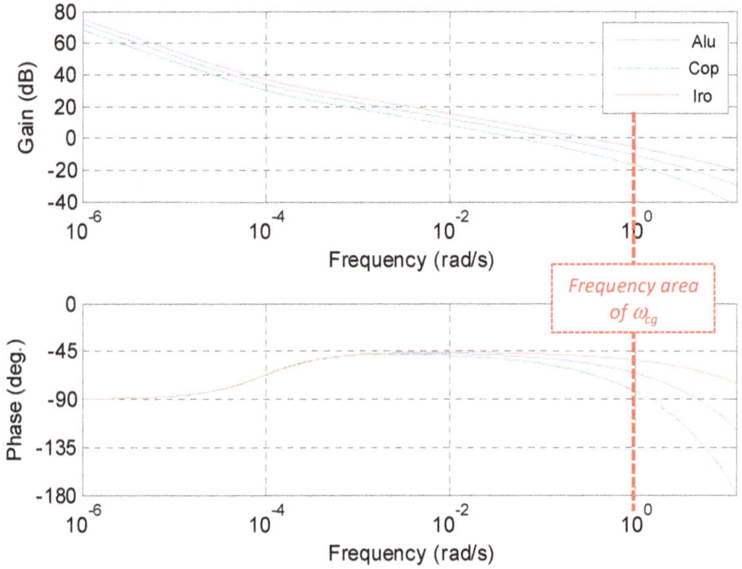

Figure 8. Bode plots of $H(x,j\omega,L)$ for aluminum (in blue), copper (in green), and iron (in red) for the second case study.

4.3.2. Synthesis Model

Based on the exact plant transfer function shown in Equation (4), the synthesis model of the plant for this second case study can be approximated by the following transfer function:

$$P_2(s) = H_0^* \frac{(1+s/\omega_L)^{0.5}}{s/\omega_L} e^{-\sqrt{\frac{s}{\omega_x}}}, \tag{26}$$

where the values of H_0^*, ω_L and ω_x remain unchanged as they were presented for the first case study (refer to Sections 4.2.2 and 4.2.3).

4.3.3. Controller Transfer Function

As already presented, the controller synthesized for the first case study will be used for this second case, as the nominal plant transfer function (e.g., the aluminum) remains the same. Thus, the controller exact transfer function is the one presented in Equation (23), whereas its rationalized form is shown in Equation (24).

4.3.4. Performance Analysis

Concerning the behavior of the second generation CRONE controller applied to a plant where the uncertainties are modeled by gain and phase variation, the plots of Figure 9 will resume the outcome.

Figure 9 shows the open-loop Black–Nichols plots (a) and the closed-loop step responses (b) for aluminum (in blue), copper (in green) and iron (in red) for the $C_R(s)$ controller.

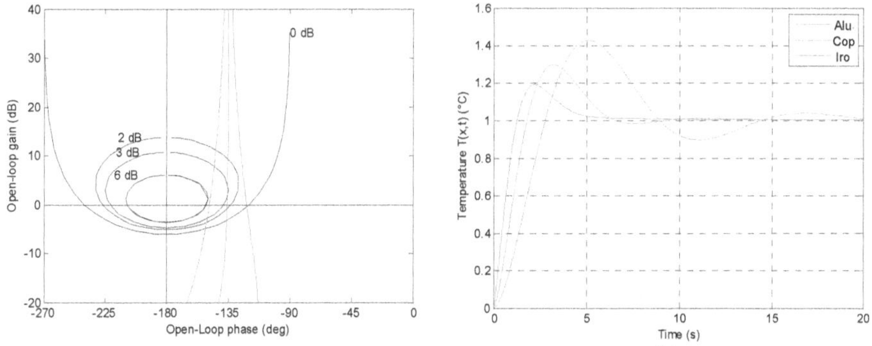

(a) (b)

Figure 9. Open-loop Black–Nichols plots (**a**) and closed-loop step responses (**b**) for aluminum (in blue), copper (in green) and iron (in red).

Concerning the robustness of this controller, one can see clearly that the three open-loop transfer functions in the Nichols diagram are no more tangent to the same contour at 3 dB. As for the closed-loop step responses, the first overshoot value and the damping ratio of the three systems are not constant (we can notice that the first overshoot varies between 19% and 42%). Thus, we can confirm that the second generation CRONE controller is no more robust whenever the conditions of the applied plant are not verified.

However, we will use this second case study to calculate the third generation CRONE controller and analyze its behavior.

5. Third CRONE Generation

5.1. Introduction

The open-loop transfer function, when using the third generation CRONE controller, is defined as being the real part of the fractional complex integrator. It is expressed as follows:

$$\beta(s) = \mathrm{Re}_{/i}\left(\frac{\omega_{cg}}{s}\right)^{n}, \tag{27}$$

where $n = a + ib \in C_i$ and $s = +j\omega \in C_j$. The real order a determines the phase placement in the Nichols chart, whereas the imaginary part b shows its angle with respect to the vertical axis, as shown in Figure 10.

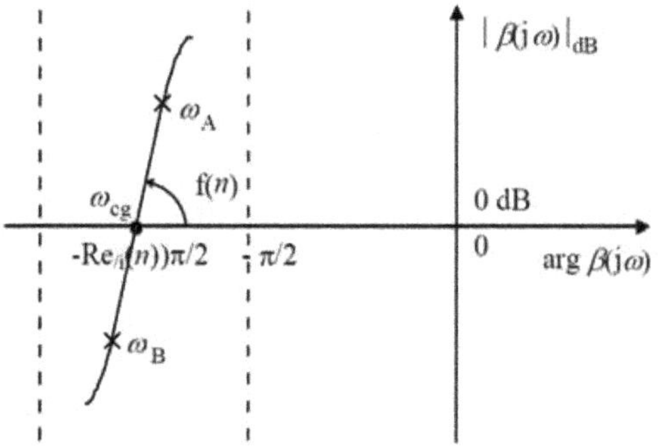

Figure 10. Open-loop behavior in the Black–Nichols diagram.

As we will have an infinite number of asymptotes, the main objective of this CRONE controller will be to optimize the parameters of the open-loop transfer function in a way that includes complex and fractional order integration on a certain frequency range, thus:

$$\beta_0(s) = \beta_l(s)\beta_m(s)\beta_h(s), \tag{28}$$

where $\beta_m(s)$ is the set of models defined within a frequency range, which allows us to write:

$$\beta_m(s) = \prod_{k=-N^-}^{N^+} C_k^{\mathrm{sign}(b_k)}\left(\alpha_k\frac{1+s/\omega_{k+1}}{1+s/\omega_k}\right)^{a_k}\left(\mathrm{Re}_{/i}\left\{\left(\alpha_k\frac{1+s/\omega_{k+1}}{1+s/\omega_k}\right)^{ib_k}\right\}\right)^{-q_k\mathrm{sign}(b_k)} \tag{29}$$

where:

$$\alpha_k = \left(\frac{\omega_{k+1}}{\omega_k}\right)^{1/2} \text{for } k \neq 0 \text{ and } \alpha_0 = \left(\frac{1+(\omega_r/\omega_0)^2}{1+(\omega_r/\omega_1)^2}\right)^{1/2} \tag{30}$$

and:

$$\beta_l(s) = C_l\left(\frac{1+s/\omega_{N^-}}{s/\omega_{N^-}}\right)^{n_l} \text{ and } \beta_h(s) = C_h\left(\frac{s}{\omega_{N^+}}+1\right)^{-n_h}, \tag{31}$$

knowing that:

$$\begin{cases} N^+,\ N^- \text{ and } q_k \in \mathbf{N}^+ \\ \omega_r, \omega_k, \omega_{k+1}, \alpha_k, C_k, C_l, C_h, a_k \text{ and } b_k \in \mathbf{R}. \end{cases} \tag{32}$$

As the calculation of all of these parameters is very difficult because of the enormous number of parameters, the CRONE toolbox has been deployed.

5.2. CRONE Toolbox

The main purpose of this toolbox is to calculate the CRONE controller transfer function based on the plant transfer function, as well as the parametric uncertainties of the system. The toolbox can specify which generation is the most suitable to answer the user guide specifications, and it can deliver the controller transfer function in a rational form.

As we are limited in space, interested authors can refer to the following references for more information concerning the toolbox and its characteristics [49,54–56].

5.3. Case Study

As already proposed, the second case study, which was proposed for the second CRONE generation, will be treated for this generation. Thus, we will be dealing with a system where both phase and gain are varying over the frequency range.

5.3.1. Plant Parameters

The plant parameters are the ones used for the second case study (refer to Section 4.3. Figure 8 showed the Bode diagram plots for the three plants. Here also, aluminum will be considered the nominal case.

5.3.2. Synthesis Model

In order to synthesize the controller transfer function, the CRONE toolbox was used for the third generation as the computation of the variable is difficult especially that the order is complex and fractional.

Hence, when using the toolbox along with the same user specifications set in the previous controller, the values of the variables of the open-loop transfer function are obtained as follow:

$$\begin{cases} \omega_0 = 0.03896\ \text{rad/s} & q_0 = 6 \\ a_0 = 1.4927 & C_0 = 13.217 \\ b_0 = -0.6496 & K = 34.11 \\ b_0' = -0.2379 & \omega_1 = 74.092\ \text{rad/s} \end{cases}. \quad (33)$$

5.3.3. Performance Analysis

Concerning the performance analysis for the third generation CRONE controller, Figure 11 represents the Bode diagrams for the rational controller $C_R(s)$ (a), the Nichols plot of the open loop (b), the Bode gain diagrams for the sensitivity functions $S(s)$ (c) and $T(s)$ (d) (as presented in system (6)), for aluminum (in blue), copper (in green) and iron (in red).

Figure 12 represents the closed step response for the temperature at location x and for time t (a), and the corresponding control signals $u(t)$ (b) for a step input of 1 °C. This will be applied for aluminum (in blue), copper (in green), and iron (in red).

As for the results, Figure 11b shows that the three Nichols plots are tangent to the same contour (3 dB) for a given gain, which may reflect the robustness of the controller. Almost the same results appear when plotting the sensitivity functions where the resonance factor is constant for the three plants (Figure 11c,d). Concerning the time domain responses, it is clear that the three plants have the same first overshoot and the same damping factor, which can prove, once again, the robustness of the third generation CRONE controller.

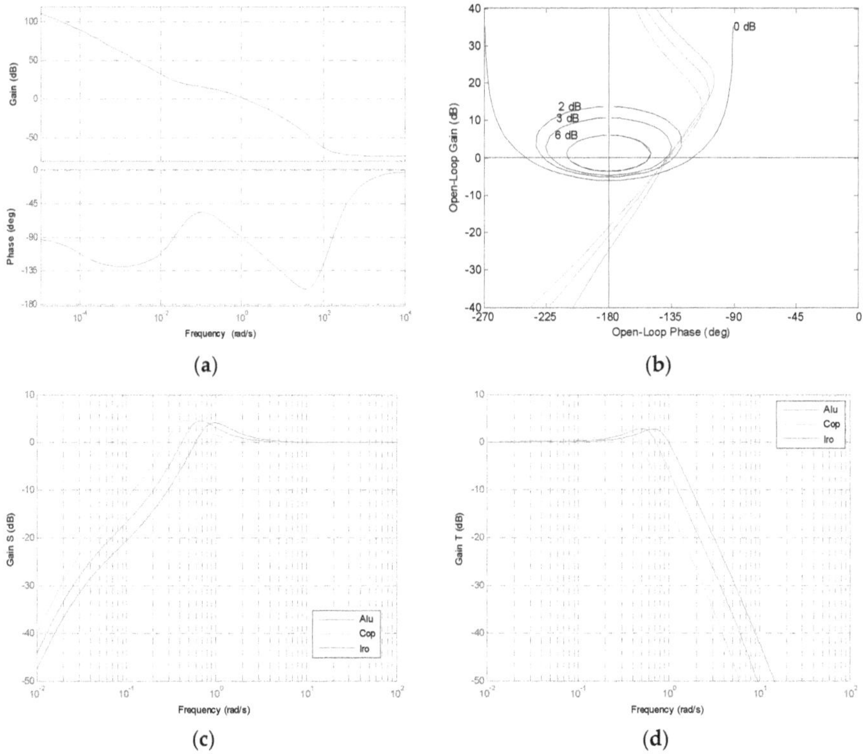

Figure 11. Frequency responses for: Bode diagrams of the controller $C_R(s)$ (**a**), the Nichols plot of the open-loop function (**b**), the sensitivity function $S(s)$, and (**c**) complementary sensitivity function $T(s)$ (**d**) for aluminum (in blue), copper (in green), and iron (in red).

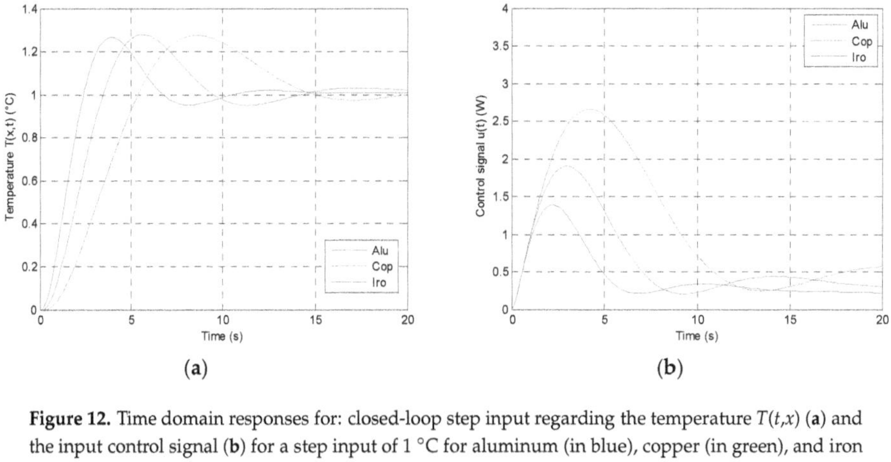

Figure 12. Time domain responses for: closed-loop step input regarding the temperature $T(t,x)$ (**a**) and the input control signal (**b**) for a step input of 1 °C for aluminum (in blue), copper (in green), and iron (in red).

6. Conclusions and Future Works

As a conclusion, the results show that the second generation CRONE controller is robust when the variations in the plant are modeled with gain changes, whereas the phase remains the same for all of the plants (even if not constant). Nevertheless, the third generation CRONE controller showed a good robustness when changing the parameters of the plant and when encountering both gain and phase variations.

As for the future works, lots of ideas come to mind in order to enrich this study. Below are some of the proposed tasks:

- Implement this system on a real test bench;
- Study the accuracy of this system when varying the position of the temperature sensors; this deviation is due involuntarily when implementing the test bench;
- Apply other regulators to control this fractional order plant as the sliding mode control (with its multiple types), H_{inf} robust control, and much more;
- Introduce some estimators to evaluate the temperature value at some location where the temperature sensor can't be placed.

Author Contributions: Xavier Moreau and Roy Abi Zeid Daou conceived the simulator of the thermal diffusive interface medium; Fady Christophy and Xavier Moreau designed the CRONE controllers and performed the simulations; Roy Abi Zeid Daou and Xavier Moreau and analyzed the data; Roy Abi Zeid Daou and Xavier Moreau wrote the paper.

Conflicts of Interest: The authors declare no conflict of interest.

References

1. Vašak, M.; Starčić, A.; Martinčević, A. Model predictive control of heating and cooling in a family house. In Proceedings of the 34th International Convention MIPRO, Opatija, Croatia, 23–27 May 2011.
2. Van Leeuwen, R.; de Wit, J.; Fink, J.; Smit, G. House thermal model parameter estimation method for Model Predictive Control applications. In Proceedings of the IEEE Eindhoven PowerTech, Eindhoven, The Netherlands, 29 June–2 July 2015.
3. Mihai, D. Fuzzy control for temperature of the driver seat in a car. In Proceedings of the 2012 International Conference on Applied and Theoretical Electricity (ICATE), Craiova, Romania, 25–27 October 2012.
4. He, B.; Liang, R.; Wu, J.; Wang, X. A Temperature Controlled System for Car Air Condition Based on Neuro-fuzzy. In Proceedings of the International Conference on Multimedia Information Networking and Security, Wuhan, China, 18–20 November 2009.
5. Erikson, B. Insulation temperature standards for industrial-control coils. *Electr. Eng.* **1944**, *63*, 546–548. [CrossRef]
6. Jones, B. Thermal Co-ordination of Motors, Control, and Their Branch Circuits on Power Supplies of 600 Volts and Less. *Trans. Am. Inst. Electr. Eng.* **1942**, *61*, 483–487. [CrossRef]
7. Zucker, M. Thermal rating of overhead line wire. *Electr. Eng.* **1943**, *62*, 501–507. [CrossRef]
8. Moore, R. The control of a thermal neutron reactor. *Proc. IEE Part II Power Eng.* **1953**, *100*, 197–198. [CrossRef]
9. Bowen, J.H. Automatic control characteristics of thermal neutron reactors. *J. Inst. Electr. Eng.* **1953**, *100*, 122–123.
10. Fink, L.H. Control of thermal environment of buried cable Systems. *Electr. Eng.* **1954**, *73*, 406–412. [CrossRef]
11. Schmill, J.V. Mathematical Solution to the Problem of the Control or the Thermal Environment or Buried Cables. *Trans. Am. Inst. Electr. Eng. Part III Power Appar. Syst.* **1960**, *79*, 175–180. [CrossRef]
12. Bhuvaneswari, T.; Yao, J.H. Automated greenhouse. In Proceedings of the IEEE International Symposium on Robotics and Manufacturing Automation, Kuala Lumpur, Malaysia, 15–16 December 2014.
13. Rodríguez-Gracia, D.; Piedra-Fernández, J.; Iribarne, L. Adaptive Domotic System in Green Buildings. In Proceedings of the 4th International Congress on Advanced Applied Informatics, Okayama, Japan, 12–16 July 2015.
14. Kumar, A. Numerical Modeling of the Thermal Boundary Layer near a Synthetic Crude Oil Plant. *J. Air Pollut. Control Assoc.* **1979**, *29*, 827–832. [CrossRef] [PubMed]

15. Van Schravendijk, B.; de Koning, W.; Nuijen, W. Modeling and control of the wafer temperatures in a diffusion furnace. *J. Appl. Phys.* **1987**, *61*, 1620–1627. [CrossRef]
16. De Waard, H.; de Koning, W. Modeling and control of diffusion and low-pressure chemical vapor deposition furnaces. *J. Appl. Phys.* **1990**, *67*, 2264–2271. [CrossRef]
17. Özişik, M.-N. *Heat Conduction*; John Wiley & Sons: Ney York, NY, USA, 1980.
18. Özişik, M.-N. *Heat Transfer, a Basic Approach*; McGraw-Hill: New York, NY, USA, 1985.
19. Trigeassou, J.-C.; Poinot, T.; Lin, J.; Oustaloup, A.; Levron, F. Modeling and identification of a non integer order system. In Proceedings of the European Control Conference, Karlsruhe, Germany, 31 August–3 September 1999.
20. Benchellal, A.; Poinot, T.; Trigeassou, J.-C. Approximation and identification of diffusive interfaces by fractional models. *Signal Proc.* **2006**, *86*, 2712–2727. [CrossRef]
21. Benchellal, A.; Poinot, T.; Trigeassou, J.-C. Fractional Modelling and Identification of a Thermal Process. *J. Vib. Control* **2008**, *14*, 1403–1414. [CrossRef]
22. Battaglia, J.; Cois, O.; Puigsegur, L.; Oustaloup, A. Solving an inverse heat conduction problem using a non-integer identified model. *Int. J. Heat Mass Transf.* **2001**, *44*, 2671–2680. [CrossRef]
23. Battaglia, J.-L.; Maachou, A.; Malti, R.; Melchior, P.; Oustaloup, A. Nonlinear heat diffusion simulation using Volterra series expansion. *Int. J. Therm. Sci.* **2013**, *71*, 80–87. [CrossRef]
24. Maachou, A.; Malti, R.; Melchior, P.; Battaglia, J.-L.; Oustaloup, A.; Hay, B. Application of fractional Volterra series for the identification of thermal diffusion in an ARMCO iron sample subject to large temperature variations. In Proceedings of the 18th IFAC World Congress (IFAC WC'11), Milano, Italy, 28 August–2 September 2011.
25. Maachou, A.; Malti, R.; Melchior, P.; Battaglia, J.-L.; Oustaloup, A.; Hay, B. Thermal identification using fractional linear models at high temperatures. In Proceedings of the 4th IFAC Workshops on Fractional Differentiation and its Applications (IFAC FDA'10), Badajoz, Spain, 18–20 October 2010.
26. Malti, R.; Sabatier, J.; Akçay, H. Thermal modeling and identification of an aluminium rod using fractional calculus. In Proceedings of the 15th IFAC Symposium on System Identification, Saint-Malo, France, 6–8 July 2009.
27. Drapaca, C.; Sivaloganathan, S. A fractional model of continuum mechanics. *J. Elast.* **2012**, *107*, 107–123. [CrossRef]
28. Sumelka, W. Thermoelasticity in the framework of the fractional continuum mechanics. *J. Therm. Stress.* **2014**, *37*, 678–706. [CrossRef]
29. Lazopoulos, K.; Lazopoulos, A. On fractional bending of beams. *Arch. Appl. Mech.* **2016**, *86*, 1133–1145. [CrossRef]
30. Bennett, S. A Brief History of Automatic Control. *IEEE Control Syst.* **1996**, *16*, 17–25. [CrossRef]
31. Bode, H. *Network Analysis and Feedback Amplifier Design*; D. Van Nostrand. Co.: New York, NY, USA, 1945.
32. Miller, K.; Ross, B. *An Introduction to the Fractional Calculus and Fractional Differential Equations*; Wiley: New York, NY, USA, 1993.
33. Samko, S.; Kilbas, A.; Marichev, O. *Fractional Integrals and Derivatives: Theory and Applications*; Gordon and Breach: Amesterdam, The Netherlands, 1993.
34. Oustaloup, A. *Etude et Réalisation d'un Systme D'asservissement D'ordre 3/2 de la Fréquence d'un Laser à Colorant Continu*; Universitu of Bordeaux: Bordeaux, France, 1975.
35. Oustaloup, A. *La Commande CRONE*; Hermes: Paris, France, 1991.
36. Oustaloup, A. *La Dérivation non Entière: Théorie, Synthèse et Applications*; Hermes: Paris, France, 1995.
37. Moreau, X.; Altet, O.; Oustaloup, A. Fractional differentiation: An example of phenomenological interpretation. In *Fractional Differentiation and Its Applications*; Ubooks Verlag: Neusäß, Germany, 2005; pp. 275–287.
38. Moreau, X.; Ramus-Serment, C.; Oustaloup, A. Fractional Differentiation in Passive Vibration Control. *J. Nonlinear Dyn.* **2002**, *29*, 343–362. [CrossRef]
39. Ortigueira, M. Introduction to fractional linear systems. Part 1. Continuous-time case. *IEE Proc. Vis. Image Signal Process.* **2000**, *147*, 62–70. [CrossRef]
40. Magin, R.; Ortigueira, M.; Podlubny, I.; Trujillo, J. On the fractional signals and systems. *Signal Process.* **2011**, *91*, 350–371. [CrossRef]
41. Machado, J.-A. Fractional Order Systems. *Nonlinear Dyn.* **2002**, *29*, 315–342. [CrossRef]

42. Ionescu, C.; Machado, J.; de Keyser, R. Modeling of the Lung Impedance Using a Fractional-Order Ladder Network With Constant Phase Elements. *IEEE Trans. Biomed. Circuits Syst.* **2011**, *5*, 83–89. [CrossRef] [PubMed]
43. Ortigueira, M.; Machado, J.-A.; Trujillo, J.; Vinagre, B. Advances in Fractional Signals and Systems. *Signal Process.* **2011**, *91*, 350–371. [CrossRef]
44. Tejado, I.; Vinagre, B.; Torres, D.; Pérez, E. Fractional disturbance observer for vibration suppression of a beam-cart system. In Proceedings of the 10th International Conference on Mechatronic and Embedded Systems and Applications, Senigallia, Italy, 10–12 September 2014.
45. Assaf, R.; Moreau, X.; Daou, R.A.Z.; Christohpy, F. Analysis of hte Fractional Order System in hte thermal diffusive interface—Part 2: Application to a finite medium. In Proceedings of the 2nd International Conference on Advances in Computational Tools for Engineering Applications, Beirut, Lebanon, 12–15 December 2012.
46. Assaf, R. Modélisation des Phénomènes de Diffusion Thermique Dans un Milieu fini Homogène en vue de L'analyse, de la Synthèse et de la Validation de Commandes Robustes. Ph.D. Thsis, Université de Bordeaux, Bordeaux, France, 2015.
47. Christophy, F.; Moreau, X.; Assaf, R.; Daou, R.A.Z. Temperature Control of a Semi Infinite Diffusive Interface Medium Using the CRONE Controller. In Proceedings of the 3rd International Conference on Control, Decision and Information Technologies (CoDIT'16), St. Julian's, Malta, 6–8 April 2016.
48. Daou, R.A.Z.; Moreau, X.; Christophy, F. Temperature Control of a finite Diffusive Interface Medium Applying CRONE Second Generation. In Proceedings of the 3rd International Conference on Advances in Computational Tools for Engineering Applications, Beirut, Lebanon, 13–15 July 2016.
49. Lanusse, P.; Malti, R.; Melchior, P. CRONE control system design toolbox for the control engineering community: Tutorial and case study. *Philos. Trans. R. Soc.* **2013**, *371*, 20120149. [CrossRef] [PubMed]
50. Oustaloup, A. *Systèmes Asservis Linéaires d'Ordre Fractionnaire*; Masson: Paris, France, 1983.
51. Oustaloup, A. *La Dérivation d'Ordre Non Entier*; Hermes: Paris, France, 1995.
52. CRONE Group. *CRONE Control Design Modul*; Bordeaux University: Bordeaux, France, 2005.
53. Daou, R.A.Z.; Moreau, X.; Christophy, F. Temperature control of a finite diffusive interface medium using the third generation CRONE controller. In Proceedings of the 20th IFAC World Congress Program, Toulouse, France, 9–14 July 2017.
54. Lanusse, P.; Nelson-Gruel, D.; Lamara, A. Toward a CRONE toolbox for the design of full MIMO controllers. In Proceedings of the 7th International Conference on Fractional Differentiation and Its Applications (IFAC-IEEE-ICFDA'16), Novi Sad, Serbia, 18–20 July 2016.
55. Oustaloup, A.; Melchior, P.; Lanusse, P.; Cois, O.; Dancla, F. The CRONE toolbox for Matlab. In Proceedings of the IEEE International Symposium on Computer-Aided Control System Design, Anchorage, AK, USA, 25–27 September 2000.
56. Malti, R.; Melchior, P.; Lanusse, P.; Oustaloup, A. Towards an object oriented CRONE toolbox for fractional differential systems. In Proceedings of the 18th IFAC World Congress, Milano, Italy, 28 August–2 September 2011.

© 2018 by the authors. Licensee MDPI, Basel, Switzerland. This article is an open access article distributed under the terms and conditions of the Creative Commons Attribution (CC BY) license (http://creativecommons.org/licenses/by/4.0/).

fractal and fractional

MDPI

Article

Dynamics and Stability Results for Hilfer Fractional Type Thermistor Problem

D. Vivek [1], K. Kanagarajan [1] and Seenith Sivasundaram [2,*]

1 Department of Mathematics, Sri Ramakrishna Mission Vidyalaya College of Arts and Science, Coimbatore 641020, India; peppyvivek@rmv.ac.in (D.V.); kanagarajank@gmail.com (K.K.)
2 Department of Mathematics, Embry-Riddle Aeronautical University, Daytona Beach, FL 32114, USA
* Correspondence: seenithi@gmail.com

Received: 22 August 2017; Accepted: 6 September 2017; Published: 9 September 2017

Abstract: In this paper, we study the dynamics and stability of thermistor problem for Hilfer fractional type. Classical fixed point theorems are utilized in deriving the results.

Keywords: nonlocal thermistor problem; Hilfer fractional derivative; existence; Ulam stability; fixed point

MSC: 26A33; 26E70; 35B09; 45M20

---

1. Introduction

Fractional differential equations (FDEs) occur in many engineering systems and scientific disciplines such as the mathematical modelling of systems and processes in the fields of physics, chemistry, aerodynamics, electrodynamics of complex medium, etc. FDEs also provide as an efficient tool for explanations of hereditary properties of different resources and processes. As a result, the meaning of the FDEs has been of great importance and attention, and one can refer to Kilbas [1], Podlubny [2] and the papers [3–9]. Recently, the Hilfer fractional derivative [10] for FDEs has become a very active area of research. R. Hilfer initiated the Hilfer fractional derivative. This is used to interpolate both the Riemann–Liouville and the Caputo fractional derivative for the theory and applications of the Hilfer fractional derivative (see, e.g., [6,10–16] and references cited therein). Analogously, we prefer the Hilfer derivative operator that interpolates both the Riemann–Liouville and the Caputo derivative.

English scientist Michael Faraday first discovered the concept of thermistors in 1833 while reporting on the semiconductor behavior of silver sulfide. From his research work, he noticed that the silver sulfides resistance decreased as the temperature increased. This later leads to the commercial production of thermistors in the 1930s when Samuel Ruben invented the first commercial thermistor. Ever since, technology has improved; this made it possible to improve manufacturing processes along with the availability of advanced quality material.

A thermistor is a thermally sensitive resistor that displays a precise and predictable change in resistance proportional to small changes in body temperature. How much its resistance will change is dependent upon its unique composition. Thermistors are part of a larger group of passive components. Unlike their active component counterparts, passive devices are incapable of providing power gain, or amplification to a circuit. Thermistors can be found everywhere in airplanes, air conditioners, in cars, computers, medical equipment, hair dryers, portable heaters, incubators, electrical outlets, refrigerators, digital thermostats, ovens, stove tops and in all kinds of appliances. Ice sensors and aircraft wings, if ice builds up on the wings, the thermistor senses this temperature drop and a heater will be activated to remove the ice. Flight tests need to be completed on a particular date, hence there may not be enough time to create a flight test technique on that date. However, it is possible to take a number of recommendations on the needs of any future flight plan to examine the nature of thermistor thermometer at high subsonic and supersonic speeds. In general, the unusual behaviour of the thermistor thermometer is caused by the possibility of vortices and an aerodynamic disturbance generating non-uniform flow, happening in the chamber with sensing element. The thermistors are small, which makes them very delicate to such effects [17,18].

A thermistor is a temperature dependent resistor and comes in two varieties, negative temperature coefficient (NTC) and positive temperature coefficient (PTC), although NTCs are most commonly used. With NTC, the resistance variation is inverse to the temperature change i.e.,: as temperature goes up, resistance goes down. NTC Thermistors are nonlinear, and their resistance decreases as temperature increases. A phenomenon called self-heating may affect the resistance of an NTC thermistor. When current flows through the NTC thermistor, it absorbs the heat causing its own temperature to rise. In [19], Khan et al. investigated the coupled p-Laplacian fractional differential equations with nonlinear boundary conditions. Wenjing Song and Wenjie Gao studied the existence of solutions for a nonlocal initial value problem to a p-Laplacian thermistor problems on time scales in [20]. Later, Moulay Rchid Sidi Ammi and Delfim F. M. Torres developed and applied a numerical method for the time-fractional nonlocal thermistor problem in [21]. They investigated the existence and uniqueness of a positive solution to generalized nonlocal thermistor problems with fractional-order derivatives in [22]. Recently, Moulay Rchid Sidi Ammi and Delfim F. M. Torres [23] discussed the existence and uniqueness results for a fractional Riemann–Liouville nonlocal thermistor problem on arbitrary time scales. Interested readers can refer to recent papers [22–26] treating a nonlocal thermistor problem.

Motivated by the aforementioned papers, we study the existence, uniqueness and Ulam–Hyers stability types of solutions for Hilfer type thermistor problem of the form

$$\begin{cases} D_{0^+}^{\alpha,\beta}u(t) = \dfrac{\lambda f(u(t))}{\left(\int_0^T f(u(x))dx\right)^2}, & t \in J := [0,T], \\ I_{0^+}^{1-\gamma}u(0) = u_0, \quad \gamma = \alpha+\beta-\alpha\beta, \end{cases} \tag{1}$$

where $D_{0^+}^{\alpha,\beta}$ is the Hilfer fractional derivative of order α and type β, $0<\alpha<1$, $0 \le \beta \le 1$ and let $J=[0,T]$, X be a Banach space, $f : J \times X \to X$ is a given continuous function. The operator $I_{0^+}^{1-\gamma}$ denotes the left-sided Riemann–Liouville fractional integral of order $1-\gamma$. Choosing λ such that $0<\lambda<\left(\frac{LT^{\alpha+1-\gamma}}{(C_1T)^2\Gamma(\alpha+1)} + \frac{2C_2^2LT^{\alpha+3-\gamma}}{(C_1T)^2\Gamma(\alpha+1)}\right)^{-1}$ is discussed in Section 4.

It is seen that (1) is equivalent to the following nonlinear integral equation

$$u(t) = \frac{u_0}{\Gamma(\gamma)}t^{\gamma-1} + \frac{\lambda}{\Gamma(\alpha)}\int_0^t (t-s)^{\alpha-1}\frac{f(u(s))}{\left(\int_0^T f(u(x))dx\right)^2}ds. \tag{2}$$

The stability of the functional equations were first introduced in a discourse conveyed in 1940 at the University of Wisconsin. The issue made by Ulam is as per the following: Under what conditions does there exist an additive mapping near an approximately additive mapping? [5,27–29]. The first reply to the topic of Ulam was given by Hyers in 1941 on account of Banach spaces. Ever since, this type of stability was known as the Ulam–Hyers stability. Rassias [29] gave a generalization of the Hyers theorem for linear mappings. Many mathematicians later extended the issue of Ulam in different ways. Recently, Ulam's problem was generalized for the stability of differential equations. A comprehensive interest was given to the study of the Ulam and Ulam–Hyers–Rassias stability of all kinds of functional equations [5,8,9,30]. An exhaustive interest was given to the investigation of the Ulam and Ulam–Hyers–Rassias stability of all kinds of functional Equation (1).

The paper is organized as follows. In Section 2, we introduce some definitions, notations, and lemmas that are used throughout the paper. In Section 3, we will prove existence and uniqueness results concerning problem (1). Section 4 is devoted to the Ulam–Hyers stabilities of problem (1).

2. Basic Concepts and Results

In this section, we introduce notations, definitions, and preliminary facts that are used throughout this paper. For more details on Hilfer fractional derivative, interested readers can refer to [6,10,12,13,15,31].

Definition 1. *Let $C[J,X]$ denote the Banach space of all continuous functions from $[0,T]$ into X with the norm*

$$\|u\|_C := \sup\{|u(t)| : t \in J\}.$$

We denote $L^1\{R_+\}$, the space of Lebesgue integrable functions on J.

By $C_\gamma[J,X]$ and $C^1_\gamma[J,X]$, we denote the weighted spaces of continuous functions defined by

$$C_\gamma[J,X] := \{f(t) : J \to X | t^\gamma f(t) \in C[J,X]\},$$

with the norm

$$\|f\|_{C_\gamma} = \|t^\gamma f(t)\|_C,$$

and

$$\|f\|_{C^n_\gamma} = \sum_{k=0}^{n-1} \left\|f^k\right\|_C + \left\|f^{(n)}\right\|_{C_\gamma}, \quad n \in N.$$

Moreover, $C^0_\gamma[J,X] := C_\gamma[J,X]$.

Now, we give some results and properties of fractional calculus.

Definition 2 ([1,16]). *The left-sided mixed Riemann–Liouville integral of order $\alpha > 0$ of a function $h \in L^1\{R_+\}$ is defined by*

$$(I^\alpha_{0^+}h)(t) = \frac{1}{\Gamma(\alpha)}\int_0^t (t-s)^{\alpha-1}h(s)ds, \quad \text{for a.e. } t \in J,$$

where $\Gamma(\cdot)$ is the (Euler's) Gamma function defined by

$$\Gamma(\xi) = \int_0^\infty t^{\xi-1}e^{-t}dt; \quad \xi > 0.$$

Notice that for all $\alpha, \alpha_1, \alpha_2 > 0$ and each $h \in C[J,X]$, we have $I^\alpha_{0^+}h \in C[J,X]$, and

$$(I^{\alpha_1}_{0^+} I^{\alpha_2}_{0^+} h)(t) = (I^{\alpha_1+\alpha_2}_{0^+} h)(t); \text{for a.e. } t \in J.$$

Definition 3 ([1,16]). *The Riemann–Liouville fractional derivative of order $\alpha \in (0,1]$ of a function $h \in L^1\{R_+\}$ is defined by*

$$\begin{aligned}(D^\alpha_{0^+}h)(t) &= \left(\frac{d}{dt}I^{1-\alpha}_{0^+}h\right)(t) \\ &= \frac{1}{\Gamma(1-\alpha)}\frac{d}{dt}\int_0^t (t-s)^{-\alpha}h(s)ds; \quad \text{for a.e. } t \in J.\end{aligned}$$

Let $\alpha \in (0,1]$, $\gamma \in [0,1)$ and $h \in C_{1-\gamma}[J,X]$. Then, the following expression leads to the left inverse operator as follows:

$$(D^{\alpha}_{0^+} I^{\alpha}_{0^+} h)(t) = h(t); \quad \text{for all } t \in (0,T].$$

Moreover, if $I^{1-\alpha}_{0^+} h \in C^1_{1-\gamma}[J,X]$, then the following composition

$$(I^{\alpha}_{0^+} D^{\alpha}_{0^+} h)(t) = h(t) - \frac{(I^{1-\alpha}_{0^+} h)(0^+)}{\Gamma(\alpha)} t^{\alpha-1}; \quad \text{for all } t \in (0,T].$$

Definition 4 ([1,16]). *The Caputo fractional derivative of order $\alpha \in (0,1]$ of a function $h \in L^1\{R_+\}$ is defined by*

$$\begin{aligned}({}^cD^{\alpha}_{0^+} h)(t) &= (I^{1-\alpha}_{0^+} \frac{d}{dt} h)(t) \\ &= \frac{1}{\Gamma(1-\alpha)} \int_0^t (t-s)^{-\alpha} \frac{d}{ds} h(s) ds; \quad \text{for a.e. } t \in J.\end{aligned}$$

In [10], Hilfer studied applications of a generalized fractional operator having the Riemann–Liouville and the Caputo derivatives as specific cases (see also [6,32]).

Definition 5 (Hilfer derivative). *Let $0 < \alpha < 1$, $0 \le \beta \le 1$, $h \in L^1\{R_+\}$, $I^{(1-\alpha)(1-\beta)}_{0^+} \in C^1_{\gamma}[J,X]$. The Hilfer fractional derivative of order α and type β of h is defined as*

$$(D^{\alpha,\beta}_{0^+} h)(t) = \left(I^{\beta(1-\alpha)}_{0^+} \frac{d}{dt} I^{(1-\alpha)(1-\beta)}_{0^+} h \right)(t); \quad \text{for a.e. } t \in J. \tag{3}$$

Properties. Let $0 < \alpha < 1$, $0 \le \beta \le 1$, $\gamma = \alpha + \beta - \alpha\beta$, and $h \in L^1\{R_+\}$.

1. The operator $(D^{\alpha,\beta}_{0^+} h)(t)$ can be written as

$$(D^{\alpha,\beta}_{0^+} h)(t) = \left(I^{\beta(1-\alpha)}_{0^+} \frac{d}{dt} I^{1-\gamma}_{0^+} h \right)(t) = \left(I^{\beta(1-\alpha)}_{0^+} D^{\gamma}_{0^+} h \right)(t); \quad \text{for a.e. } t \in J.$$

   Moreover, the parameter γ satisfies

$$0 < \gamma \le 1, \quad \gamma \ge \alpha, \quad \gamma > \beta, \quad 1-\gamma < 1 - \beta(1-\alpha).$$

2. The generalization (3) for $\beta = 0$ coincides with the Riemann–Liouville derivative and for $\beta = 1$ with the Caputo derivative

$$D^{\alpha,0}_{0^+} = D^{\alpha}_{0^+}, \quad \text{and} \quad D^{\alpha,1}_{0^+} = {}^cD^{\alpha}_{0^+}.$$

3. If $D^{\beta(1-\alpha)}_{0^+} h$ exists and in $L^1\{R_+\}$, then

$$(D^{\alpha,\beta}_{0^+} I^{\alpha}_{0^+} h)(t) = \left(I^{\beta(1-\alpha)}_{0^+} D^{\beta(1-\alpha)}_{0^+} h \right)(t); \quad \text{for a.e. } t \in J.$$

   Furthermore, if $h \in C_{\gamma}[J,X]$ and $I^{1-\beta(1-\alpha)}_{0^+} h \in C^1_{\gamma}[J,X]$, then

$$(D^{\alpha,\beta}_{0^+} I^{\alpha}_{0^+} h)(t) = h(t); \quad \text{for a.e. } t \in J.$$

4. If $D_{0^+}^{\gamma} h$ exists and in $L^1\{R_+\}$, then

$$\left(I_{0^+}^{\alpha} D_{0^+}^{\alpha,\beta} h\right)(t) = \left(I_{0^+}^{\gamma} D_{0^+}^{\gamma} h\right)(t) = h(t) - \frac{I_{0^+}^{1-\gamma} h(0^+)}{\Gamma(\gamma)} t^{\gamma-1}; \quad \text{for a.e. } t \in J.$$

In order to solve our problem, the following spaces are presented

$$C_{1-\gamma}^{\alpha,\beta}[J,X] = \left\{ f \in C_{1-\gamma}[J,X], D_{0^+}^{\alpha,\beta} f \in C_{1-\gamma}[J,X] \right\},$$

and

$$C_{1-\gamma}^{\gamma}[J,X] = \left\{ f \in C_{1-\gamma}[J,X], D_{0^+}^{\gamma} f \in C_{1-\gamma}[J,X] \right\}.$$

It is obvious that

$$C_{1-\gamma}^{\gamma}[J,X] \subset C_{1-\gamma}^{\alpha,\beta}[J,X].$$

Corollary 1 ([31]). *Let $h \in C_{1-\gamma}[J,X]$. Then, the linear problem*

$$D_{0^+}^{\alpha,\beta} x(t) = h(t), \quad t \in J = [0,T],$$
$$I_{0^+}^{1-\gamma} x(0) = x_0, \quad \gamma = \alpha + \beta - \alpha\beta,$$

has a unique solution $x \in L^1\{R_+\}$ given by

$$x(t) = \frac{x_0}{\Gamma(\gamma)} t^{\gamma-1} + \frac{1}{\Gamma(\alpha)} \int_0^t (t-s)^{\alpha-1} h(s) ds.$$

From the above corollary, we conclude the following lemma.

Lemma 1. *Let $f : J \times X \to X$ be a function such that $f \in C_{1-\gamma}[J,X]$. Then, problem* (1) *is equivalent to the problem of the solutions of the integral Equation* (2).

Theorem 1 (Schauder fixed point theorem [31,33]). *Let B be closed, convex and nonempty subset of a Banach space E. Let $P : B \to B$ be a continuous mapping such that $P(B)$ is a relatively compact subset of E. Then, P has at least one fixed point in B.*

Now, we study the Ulam stability, and we adopt the definitions in [4,30,34] of the Ulam–Hyers stability, generalized Ulam–Hyers stability, Ulam–Hyers–Rassias stability and generalized Ulam–Hyers–Rassias stability.

Consider the following Hilfer type termistor problem

$$D_{0^+}^{\alpha,\beta} u(t) = \frac{\lambda f(u(t))}{\left(\int_0^T f(u(x)) dx\right)^2}, \quad t \in J := [0,T], \tag{4}$$

and the following fractional inequalities:

$$\left| D_{0+}^{\alpha,\beta} z(t) - \frac{\lambda f(z(t))}{\left(\int_0^T f(z(x))dx\right)^2} \right| \leq \epsilon, \quad t \in J, \tag{5}$$

$$\left| D_{0+}^{\alpha,\beta} z(t) - \frac{\lambda f(z(t))}{\left(\int_0^T f(z(x))dx\right)^2} \right| \leq \epsilon\varphi(t), \quad t \in J, \tag{6}$$

$$\left| D_{0+}^{\alpha,\beta} z(t) - \frac{\lambda f(z(t))}{\left(\int_0^T f(z(x))dx\right)^2} \right| \leq \varphi(t), \quad t \in J. \tag{7}$$

Definition 6. *Equation (4) is Ulam–Hyers stable if there exists a real number $C_f > 0$ such that, for each $\epsilon > 0$ and for each solution $z \in C_{1-\gamma}^{\gamma}[J,X]$ of Inequality (5), there exists a solution $u \in C_{1-\gamma}^{\gamma}[J,X]$ of Equation (4) with*

$$|z(t) - u(t)| \leq C_f \epsilon, \quad t \in J.$$

Definition 7. *Equation (4) is generalized Ulam–Hyers stable if there exists $\psi_f \in C([0,\infty),[0,\infty)), \psi_f(0) = 0$ such that, for each solution $z \in C_{1-\gamma}^{\gamma}[J,X]$ of Inequality (5), there exists a solution $u \in C_{1-\gamma}^{\gamma}[J,X]$ of Equation (4) with*

$$|z(t) - u(t)| \leq \psi_f \epsilon, \quad t \in J.$$

Definition 8. *Equation (4) is Ulam–Hyers–Rassias stable with respect to $\varphi \in C_{1-\gamma}[J,X]$ if there exists a real number $C_f > 0$ such that, for each $\epsilon > 0$ and for each solution $z \in C_{1-\gamma}^{\gamma}[J,X]$ of Inequality (6), there exists a solution $u \in C_{1-\gamma}^{\gamma}[J,X]$ of Equation (4) with*

$$|z(t) - u(t)| \leq C_f \epsilon \varphi(t), \quad t \in J.$$

Definition 9. *Equation (4) is generalized Ulam–Hyers–Rassias stable with respect to $\varphi \in C_{1-\gamma}[J,X]$ if there exists a real number $C_{f,\varphi} > 0$ such that, for each solution $z \in C_{1-\gamma}^{\gamma}[J,X]$ of Inequality (7), there exists a solution $u \in C_{1-\gamma}^{\gamma}[J,X]$ of Equation (4) with*

$$|z(t) - u(t)| \leq C_{f,\varphi} \varphi(t), \quad t \in J.$$

Remark 1. *A function $z \in C_{1-\gamma}^{\gamma}[J,X]$ is a solution of Inequality (5) if and only if there exist a function $g \in C_{1-\gamma}^{\gamma}[J,X]$ (which depends on solution z) such that*

1. $|g(t)| \leq \epsilon, \quad \forall\, t \in J.$
2. $D_{0+}^{\alpha,\beta} z(t) = \frac{\lambda f(z(t))}{\left(\int_0^T f(z(x))dx\right)^2} + g(t), \quad t \in J.$

Remark 2. *It is clear that:*

1. *Definition 6$\Rightarrow$ Definition 7.*
2. *Definition 8$\Rightarrow$ Definition 9.*
3. *Definition 8 for $\varphi(t) = 1 \Rightarrow$ Definition 6.*

Lemma 2 ([3]). *Let $v : [0,T] \to [0,\infty)$ be a real function and $w(\cdot)$ is a nonnegative, locally integrable function on $[0,T]$ and there are constants $a > 0$ and $0 < \alpha < 1$ such that*

$$v(t) \leq w(t) + a\int_0^t \frac{v(s)}{(t-s)^{\alpha}}ds.$$

Then, there exists a constant $K = K(\alpha)$ such that

$$v(t) \leq w(t) + Ka\int_0^t \frac{w(s)}{(t-s)^{\alpha}}ds,$$

for every $t \in [0,T]$.

3. Existence Results

The following existence result for Hilfer type thermistor problem (1) is based on Schauder's fixed point theorem. Let us consider the following assumptions:

Assumption 1. *Function $f : J \times X \to X$ of problem* (1) *is Lipschitz continuous with Lipschitz constant L such that $c_1 \leq f(u) \leq c_2$, with c_1 and c_2 two positive constants.*

Assumption 2. *There exists an increasing function $\varphi \in C_{1-\gamma}[J,X]$ and there exists $\lambda_{\varphi} > 0$ such that, for any $t \in J$,*

$$I_{0^+}^{\alpha}\varphi(t) \leq \lambda_{\varphi}\varphi(t).$$

Our main result may be presented as the following theorem.

Theorem 2 (existence). *Under the above Assumption 1, problem* (1) *has at least one solution $u \in X$ for all $\lambda > 0$.*

Proof. Consider the operator $P : C_{1-\gamma}[J,X] \to C_{1-\gamma}[J,X]$ is defined by

$$(Pu)(t) = \frac{u_0}{\Gamma(\gamma)}t^{\gamma-1} + \frac{\lambda}{\Gamma(\alpha)}\int_0^t (t-s)^{\alpha-1}\frac{f(u(s))}{\left(\int_0^T f(u(x))dx\right)^2}ds. \tag{8}$$

Clearly, the fixed points of P are solutions to (1). The proof will be given in several steps.

Step 1: The operator P is continuous. Let u_n be a sequence such that $u_n \to u$ in $C_{1-\gamma}[J,X]$. Then, for each $t \in J$,

$$\begin{aligned}
&\left|t^{1-\gamma}\left((Pu_n)(t) - (Pu)(t)\right)\right| \\
&\leq \tfrac{\lambda t^{1-\gamma}}{\Gamma(\alpha)}\int_0^t (t-s)^{\alpha-1}\left|\frac{f(u_n(s))}{\left(\int_0^T f(u_n(x))dx\right)^2} - \frac{f(u(s))}{\left(\int_0^T f(u(x))dx\right)^2}\right|ds \\
&\leq \tfrac{\lambda t^{1-\gamma}}{\Gamma(\alpha)}\int_0^t (t-s)^{\alpha-1}\left|\frac{1}{\left(\int_0^T f(u_n(x))dx\right)^2}\left(f(u_n(s)) - f(u(s))\right)\right. \\
&\quad\left. + f(u(s))\left(\frac{1}{\left(\int_0^T f(u_n(x))dx\right)^2} - \frac{1}{\left(\int_0^T f(u(x))dx\right)^2}\right)\right| \\
&\leq \tfrac{\lambda t^{1-\gamma}}{\Gamma(\alpha)}\int_0^t (t-s)^{\alpha-1}\frac{1}{\left(\int_0^T f(u_n(x))dx\right)^2}\left|f(u_n(s)) - f(u(s))\right|ds \\
&\quad + \tfrac{\lambda t^{1-\gamma}}{\Gamma(\alpha)}\int_0^t (t-s)^{\alpha-1}\left|f(u(s))\right|\left|\frac{1}{\left(\int_0^T f(u_n(x))dx\right)^2} - \frac{1}{\left(\int_0^T f(u(x))dx\right)^2}\right|ds \leq I_1 + I_2,
\end{aligned} \tag{9}$$

where

$$I_1 = \frac{\lambda t^{1-\gamma}}{\Gamma(\alpha)} \int_0^t (t-s)^{\alpha-1} \frac{1}{\left(\int_0^T f(u_n(x))dx\right)^2} \left| f(u_n(s)) - f(u(s)) \right| ds,$$

$$I_2 = \frac{\lambda t^{1-\gamma}}{\Gamma(\alpha)} \int_0^t (t-s)^{\alpha-1} \left| f(u(s)) \right| \left| \frac{1}{\left(\int_0^T f(u_n(x))dx\right)^2} - \frac{1}{\left(\int_0^T f(u(x))dx\right)^2} \right| ds.$$

We estimate I_1 and I_2 terms separately. By Assumption 1, we have

$$\begin{aligned}
I_1 &\leq \frac{\lambda t^{1-\gamma}}{\Gamma(\alpha)} \int_0^t (t-s)^{\alpha-1} \frac{1}{\left(\int_0^T f(u_n(x))dx\right)^2} \left| f(u_n(s)) - f(u(s)) \right| ds \\
&\leq \frac{\lambda t^{1-\gamma}}{(c_1 T)^2 \Gamma(\alpha)} \int_0^t (t-s)^{\alpha-1} \left| f(u_n(s)) - f(u(s)) \right| ds \\
&\leq \frac{L\lambda t^{1-\gamma}}{(c_1 T)^2 \Gamma(\alpha)} \| u_n - u \|_{C_{1-\gamma}} \int_0^t (t-s)^{\alpha-1} ds \\
&\leq \frac{L\lambda T^{\alpha+1-\gamma}}{(c_1 T)^2 \Gamma(\alpha+1)} \| u_n - u \|_{C_{1-\gamma}}.
\end{aligned}$$

Then,

$$I_1 \leq \frac{L\lambda T^{\alpha+1-\gamma}}{(c_1 T)^2 \Gamma(\alpha+1)} \| u_n - u \|_{C_{1-\gamma}}, \tag{10}$$

$$\begin{aligned}
I_2 &= \frac{\lambda t^{1-\gamma}}{\Gamma(\alpha)} \int_0^t (t-s)^{\alpha-1} \left| f(u(s)) \right| \left| \frac{1}{\left(\int_0^T f(u_n(x))dx\right)^2} - \frac{1}{\left(\int_0^T f(u(x))dx\right)^2} \right| ds \\
&\leq \frac{\lambda t^{1-\gamma}}{\Gamma(\alpha)} \int_0^t (t-s)^{\alpha-1} \left| f(u(s)) \right| \frac{\left| \left(\int_0^T f(u_n(x))dx\right)^2 - \left(\int_0^T f(u(x))dx\right)^2 \right|}{\left(\int_0^T f(u_n(x))dx\right)^2 \left(\int_0^T f(u(x))dx\right)^2} ds \\
&\leq \frac{\lambda t^{1-\gamma} c_2}{(c_1 T)^4 \Gamma(\alpha)} \int_0^t (t-s)^{\alpha-1} \left| \left(\int_0^T f(u_n(x))dx\right)^2 - \left(\int_0^T f(u(x))dx\right)^2 \right| ds \\
&\leq \frac{\lambda t^{1-\gamma} c_2}{(c_1 T)^4 \Gamma(\alpha)} \int_0^t (t-s)^{\alpha-1} \left| \left(\int_0^T (f(u_n(x)) - f(u(x)))dx\right) \left(\int_0^T (f(u_n(x)) + f(u(x)))dx\right) \right| ds \\
&\leq \frac{2\lambda c_2^2 T t^{1-\gamma}}{(c_1 T)^4 \Gamma(\alpha)} \int_0^t (t-s)^{\alpha-1} \left(\int_0^T \left| f(u_n(x)) - f(u(x)) \right| dx \right) ds \\
&\leq \frac{2\lambda c_2^2 T L t^{1-\gamma}}{(c_1 T)^4 \Gamma(\alpha)} \int_0^t (t-s)^{\alpha-1} \left(\int_0^T \left| u_n(x) - u(x) \right| dx \right) ds \\
&\leq \frac{2\lambda c_2^2 L T^2 t^{1-\gamma}}{(c_1 T)^4 \Gamma(\alpha)} \| u_n - u \|_{C_{1-\gamma}} \int_0^t (t-s)^{\alpha-1} ds \\
&\leq \frac{2\lambda c_2^2 L T^{\alpha+3-\gamma}}{(c_1 T)^4 \Gamma(\alpha+1)} \| u_n - u \|_{C_{1-\gamma}}.
\end{aligned}$$

It follows that

$$I_2 \leq \frac{2\lambda c_2^2 L T^{\alpha+3-\gamma}}{(c_1 T)^4 \Gamma(\alpha+1)} \| u_n - u \|_{C_{1-\gamma}}. \tag{11}$$

To substitute (10) and (11) into (9), we have

$$\begin{aligned}\left|t^{1-\gamma}\left((Pu_n)(t)-(Pu)(t)\right)\right| &\le I_1+I_2\\ &\le \left(\frac{L\lambda T^{\alpha+1-\gamma}}{(c_1T)^2\Gamma(\alpha+1)}+\frac{2\lambda c_2^2LT^{\alpha+3-\gamma}}{(c_1T)^4\Gamma(\alpha+1)}\right)\|u_n-u\|_{C_{1-\gamma}}.\end{aligned}$$

Then,

$$\|Pu_n-Pu\|_{C_{1-\gamma}}\le\left(\frac{L\lambda T^{\alpha+1-\gamma}}{(c_1T)^2\Gamma(\alpha+1)}+\frac{2\lambda c_2^2LT^{\alpha+3-\gamma}}{(c_1T)^4\Gamma(\alpha+1)}\right)\|u_n-u\|_{C_{1-\gamma}}.$$

Here, independently of λ, the right-hand side of the above inequality converges to zero as $u_n\to u$. Therefore, $Pu_n\to Pu$. This proves the continuity of P.

Step 2: The operator P maps bounded sets into bounded sets in $C_{1-\gamma}[J,X]$.

Indeed, it is enough to show that, for $r>0$, there exists a positive constant l such that $u\in B_r\left\{u\in C_{1-\gamma}[J,X]:\|u\|\le r\right\}$, we have $\|(Pu)\|_{C_{1-\gamma}}\le l$. Set $M=\sup_{B_r}\frac{f}{(c_1T)^2}$:

$$\begin{aligned}\left|t^{1-\gamma}(Pu)(t)\right| &\le \frac{|u_0|}{\Gamma(\gamma)}+\frac{\lambda t^{1-\gamma}}{\Gamma(\alpha)}\int_0^t(t-s)^{\alpha-1}\frac{|f(u(s))|}{\left(\int_0^Tf(u(x))dx\right)^2}ds\\ &\le \frac{|u_0|}{\Gamma(\gamma)}+\frac{\lambda Mt^{1-\gamma}}{\Gamma(\alpha)}\int_0^t(t-s)^{\alpha-1}ds\\ &\le \frac{|u_0|}{\Gamma(\gamma)}+\frac{\lambda MT^{1-\gamma+\alpha}}{\Gamma(\alpha)}.\end{aligned}$$

Thus,

$$\|Pu\|_{C_{1-\gamma}}\le\frac{|u_0|}{\Gamma(\gamma)}+\frac{\lambda MT^{1-\gamma+\alpha}}{\Gamma(\alpha)}:\le l.$$

Step 3: P maps bounded sets into equicontinuous set of $C_{1-\gamma}[J,X]$.

Let $t_1,t_2\in J, t_1<t_2$, B_r be a bounded set of $C_{1-\gamma}[J,X]$ and $u\in B_r$. Then,

$$\begin{aligned}&\left|t_2^{1-\gamma}(Pu)(t_2)-t_1^{1-\gamma}(Pu)(t_1)\right|\\ &\le\frac{\lambda}{\Gamma(\alpha)}\left|t_2^{1-\gamma}\int_0^{t_2}(t_2-s)^{\alpha-1}\frac{f(u(s))}{\left(\int_0^Tf(u(x))dx\right)^2}ds-t_1^{1-\gamma}\int_0^{t_1}(t_1-s)^{\alpha-1}\frac{f(u(s))}{\left(\int_0^Tf(u(x))dx\right)^2}ds\right|\\ &\le\frac{\lambda t_2^{1-\gamma}}{\Gamma(\alpha)}\int_{t_1}^{t_2}(t_2-s)^{\alpha-1}\frac{|f(u(s))|}{\left(\int_0^Tf(u(x))dx\right)^2}ds\\ &\quad+\frac{\lambda}{\Gamma(\alpha)}\int_0^{t_2}\left|t_2^{1-\gamma}(t_2-s)^{\alpha-1}-t_1^{1-\gamma}(t_1-s)^{\alpha-1}\right|\frac{|f(u(s))|}{\left(\int_0^Tf(u(x))dx\right)^2}ds\\ &\le\frac{\lambda c_2t_2^{1-\gamma}}{(c_1T)^2\Gamma(\alpha)}\int_{t_1}^{t_2}(t_2-s)^{\alpha-1}ds+\frac{\lambda c_2}{(c_1T)^2\Gamma(\alpha)}\int_0^{t_2}\left|t_2^{1-\gamma}(t_2-s)^{\alpha-1}-t_1^{1-\gamma}(t_1-s)^{\alpha-1}\right|ds\\ &\le\frac{\lambda c_2t_2^{1-\gamma}}{(c_1T)^2\Gamma(\alpha+1)}(t_2-t_1)^{1-\alpha}+\frac{\lambda c_2}{(c_1T)^2\Gamma(\alpha)}\int_0^{t_2}\left|t_2^{1-\gamma}(t_2-s)^{\alpha-1}-t_1^{1-\gamma}(t_1-s)^{\alpha-1}\right|ds.\end{aligned}$$

Beacause the right-hand side of the above inequality does not depend on u and tends to zero when $t_2 \to t_1$, we conclude that $P(\overline{B_r})$ is relatively compact. Hence, B is compact by the Arzela–Ascoli theorem. Consequently, since P is continuous, it follows by Theorem 1 that problem (1) has a solution. The proof is completed. □

4. The Ulam–Hyers–Rassias Stability

In this section, we investigate generalized Ulam–Hyers–Rassias stability for problem (1). The stability results are based on the Banach contraction principle.

Lemma 3 (Uniqueness). *Assume that the Assumption 1 is hold. If*

$$\left(\frac{L\lambda T^{\alpha+1-\gamma}}{(c_1T)^2\Gamma(\alpha+1)}+\frac{2\lambda c_2^2LT^{\alpha+3-\gamma}}{(c_1T)^2\Gamma(\alpha+1)}\right)<1, \tag{12}$$

then problem (1) *has a unique solution.*

Proof. Consider the operator $P: C_{1-\gamma}[J,X]\to C_{1-\gamma}[J,X]$:

$$(Pu)(t)=\frac{u_0}{\Gamma(\gamma)}t^{\gamma-1}+\frac{\lambda}{\Gamma(\alpha)}\int_0^t(t-s)^{\alpha-1}\frac{f(u(s))}{\left(\int_0^T f(u(x))dx\right)^2}ds. \tag{13}$$

It is clear that the fixed points of P are solutions of problem (1).
Letting $u,v\in C_{1-\gamma}[J,X]$ and $t\in J$, then we have

$$\begin{aligned}\left|t^{1-\gamma}\left((Pv)(t)-(Pu)(t)\right)\right| &\leq \frac{\lambda t^{1-\gamma}}{\Gamma(\alpha)}\int_0^t(t-s)^{\alpha-1}\left|\frac{f(v(s))}{\left(\int_0^T f(v(x))dx\right)^2}-\frac{f(u(s))}{\left(\int_0^T f(u(x))dx\right)^2}\right|ds\\ &\leq\left(\frac{L\lambda T^{\alpha+1-\gamma}}{(c_1T)^2\Gamma(\alpha+1)}+\frac{2\lambda c_2^2LT^{\alpha+3-\gamma}}{(c_1T)^2\Gamma(\alpha+1)}\right)\|v-u\|_{C_{1-\gamma}}.\end{aligned}$$

Then,

$$\|Pv-Pu\|_{C_{1-\gamma}}\leq\left(\frac{L\lambda T^{\alpha+1-\gamma}}{(c_1T)^2\Gamma(\alpha+1)}+\frac{2\lambda c_2^2LT^{\alpha+3-\gamma}}{(c_1T)^2\Gamma(\alpha+1)}\right)\|v-u\|_{C_{1-\gamma}}.$$

Choosing λ such that $0<\lambda<\left(\frac{LT^{\alpha+1-\gamma}}{(C_1T)^2\Gamma(\alpha+1)}+\frac{2C_2^2LT^{\alpha+3-\gamma}}{(C_1T)^2\Gamma(\alpha+1)}\right)^{-1}$, the map $P: C_{1-\gamma}[J,X]\to C_{1-\gamma}[J,X]$ is a contraction. From (12), it follows that P has a unique fixed point, which is a solution of problem (1). □

Theorem 3. *In Assumption 1 and* (12), *problem* (1) *is Ulam–Hyers stable.*

Proof. Let $\epsilon>0$ and let $z\in C_{1-\gamma}^{\gamma}[J,X]$ be a function that satisfies Inequality (5) and let $u\in C_{1-\gamma}^{\gamma}[J,X]$ be the unique solution of the following Hilfer type thermistor problem

$$\begin{aligned}D_{0^+}^{\alpha,\beta}u(t)&=\frac{\lambda f(u(t))}{\left(\int_0^T f(u(x))dx\right)^2},\quad t\in J:=[0,T],\\ I_{0^+}^{1-\gamma}u(t)&=I_{0^+}^{1-\gamma}z(t)=u_0,\end{aligned}$$

where $\alpha \in (0,1)$, $\beta \in [0,1]$. From Lemma 1, we have

$$u(t) = \frac{u_0}{\Gamma(\gamma)} t^{\gamma-1} + \frac{\lambda}{\Gamma(\alpha)} \int_0^t (t-s)^{\alpha-1} \frac{f(u(s))}{\left(\int_0^T f(u(x))dx\right)^2} ds.$$

By integration of (5), we obtain

$$\left| z(t) - \frac{u_0}{\Gamma(\gamma)} t^{\gamma-1} - \frac{\lambda}{\Gamma(\alpha)} \int_0^t (t-s)^{\alpha-1} \frac{f(z(s))}{\left(\int_0^T f(z(x))dx\right)^2} ds \right| \leq \frac{\epsilon T^\alpha}{\Gamma(\alpha+1)}, \tag{14}$$

for all $t \in J$. From the above, it follows:

$$\begin{aligned}
&|z(t) - u(t)| \\
&\leq \left| z(t) - \frac{u_0}{\Gamma(\gamma)} t^{\gamma-1} - \frac{\lambda}{\Gamma(\alpha)} \int_0^t (t-s)^{\alpha-1} \frac{f(z(s))}{\left(\int_0^T f(z(x))dx\right)^2} ds \right| \\
&\quad + \frac{\lambda}{\Gamma(\alpha)} \int_0^t (t-s)^{\alpha-1} \left| \frac{f(z(s))}{\left(\int_0^T f(z(x))dx\right)^2} - \frac{f(u(s))}{\left(\int_0^T f(u(x))dx\right)^2} \right| ds \\
&\leq \frac{\epsilon T^\alpha}{\Gamma(\alpha+1)} + \frac{\lambda}{\Gamma(\alpha)} \int_0^t (t-s)^{\alpha-1} \frac{1}{\left(\int_0^T f(z(x))dx\right)^2} |f(z(s)) - f(u(s))| ds \\
&\quad + \frac{\lambda}{\Gamma(\alpha)} \int_0^t (t-s)^{\alpha-1} |f(u(s))| \left| \frac{1}{\left(\int_0^T f(z(x))dx\right)^2} - \frac{1}{\left(\int_0^T f(u(x))dx\right)^2} \right| ds.
\end{aligned} \tag{15}$$

For computational convenience, we set

$$K_1 = \frac{\lambda}{\Gamma(\alpha)} \int_0^t (t-s)^{\alpha-1} \frac{1}{\left(\int_0^T f(z(x))dx\right)^2} |f(z(s)) - f(u(s))| ds,$$

$$K_2 = \frac{\lambda}{\Gamma(\alpha)} \int_0^t (t-s)^{\alpha-1} |f(u(s))| \left| \frac{1}{\left(\int_0^T f(z(x))dx\right)^2} - \frac{1}{\left(\int_0^T f(u(x))dx\right)^2} \right| ds.$$

We estimate K_1, K_2 terms separately. By Assumption 1, we have

$$\begin{aligned}
K_1 &\leq \frac{\lambda}{\Gamma(\alpha)} \int_0^t (t-s)^{\alpha-1} \frac{1}{\left(\int_0^T f(z(x))dx\right)^2} |f(z(s)) - f(u(s))| ds \\
&\leq \frac{\lambda}{(c_1 T)^2 \Gamma(\alpha)} \int_0^t (t-s)^{\alpha-1} |f(z(s)) - f(u(s))| ds \\
&\leq \frac{\lambda L}{(c_1 T)^2 \Gamma(\alpha)} \int_0^t (t-s)^{\alpha-1} |z(s) - u(s)| ds,
\end{aligned} \tag{16}$$

$$
\begin{aligned}
K_2 &\le \frac{\lambda}{\Gamma(\alpha)} \int_0^t (t-s)^{\alpha-1} |f(u(s))| \frac{\left|\left(\int_0^T f(z(x))dx\right)^2 - \left(\int_0^T f(u(x))dx\right)^2\right|}{\left(\int_0^T f(z(x))dx\right)^2 \left(\int_0^T f(u(x))dx\right)^2} ds \\
&\le \frac{2\lambda c_2^2 TL}{(c_1T)^4\Gamma(\alpha)} \int_0^t (t-s)^{\alpha-1} \left(\int_0^T |z(x)-u(x)|\, dx\right) ds \\
&\le \frac{2\lambda c_2^2 TL}{(c_1T)^4\Gamma(\alpha)} \|z-u\|_{C_{1-\gamma}} \int_0^t (t-s)^{\alpha-1} ds \\
&\le \frac{2\lambda c_2^2 TL}{(c_1T)^4\Gamma(\alpha)} \int_0^t (t-s)^{\alpha-1} |z(s)-u(s)|\, ds.
\end{aligned} \tag{17}
$$

To substitute (16) and (17) into (15), we get

$$
\begin{aligned}
|z(t)-u(t)| &\le \frac{\epsilon T^\alpha}{\Gamma(\alpha+1)} + \frac{\lambda L}{(c_1T)^2\Gamma(\alpha)} \int_0^t (t-s)^{\alpha-1} |z(s)-u(s)|\, ds \\
&\quad + \frac{2\lambda c_2^2 T^2 L}{(c_1T)^4\Gamma(\alpha)} \int_0^t (t-s)^{\alpha-1} |z(s)-u(s)|\, ds \\
&\le \frac{\epsilon T^\alpha}{\Gamma(\alpha+1)} + \left(\frac{\lambda L}{(c_1T)^2} + \frac{2\lambda c_2^2 T^2 L}{(c_1T)^4}\right) \frac{1}{\Gamma(\alpha)} \int_0^t (t-s)^{\alpha-1} |z(s)-u(s)|\, ds,
\end{aligned}
$$

and, to apply Lemma 2, we have

$$
|z(t)-u(t)| \le \frac{T^\alpha}{\Gamma(\alpha+1)} \left[1 + \frac{\nu T^\alpha}{\Gamma(\alpha+1)} \left(\frac{\lambda L}{(c_1T)^2} + \frac{2\lambda c_2^2 T^2 L}{(c_1T)^4}\right)\right] \epsilon := C_f \epsilon,
$$

where $\nu = \nu(\alpha)$ is a constant, which completes the proof of the theorem. Moreover, if we set $\psi(\epsilon) = C_f\epsilon$; $\psi(0) = 0$, then problem (1) is generalized Ulam–Hyers stable. □

Theorem 4. *In Assumptions 1, 2 and* (12), *problem* (1) *is Ulam–Hyers–Rassias stable.*

Proof. Let $z \in C_{1-\gamma}^{\gamma}[J, X]$ be solution of Inequality (6) and let $z \in C_{1-\gamma}^{\gamma}[J, X]$ be the unique solution of the following Hilfer type thermistor problem

$$
\begin{aligned}
D_{0^+}^{\alpha,\beta} u(t) &= \frac{\lambda f(u(t))}{\left(\int_0^T f(u(x))dx\right)^2}, \quad t \in J := [0, T], \\
I_{0^+}^{1-\gamma} u(t) &= I_{0^+}^{1-\gamma} z(t) = u_0,
\end{aligned}
$$

where $\alpha \in (0,1)$, $\beta \in [0,1]$. From Lemma 1, we have

$$
u(t) = \frac{u_0}{\Gamma(\gamma)} t^{\gamma-1} + \frac{\lambda}{\Gamma(\alpha)} \int_0^t (t-s)^{\alpha-1} \frac{f(u(s))}{\left(\int_0^T f(u(x))dx\right)^2} ds.
$$

By integration of (6) and Assumption 2, we obtain

$$
\left| z(t) - \frac{u_0}{\Gamma(\gamma)} t^{\gamma-1} - \frac{\lambda}{\Gamma(\alpha)} \int_0^t (t-s)^{\alpha-1} \frac{f(z(s))}{\left(\int_0^T f(z(x))dx\right)^2} ds \right| \le \epsilon \lambda_\varphi \varphi(t), \tag{18}
$$

for all $t \in J$. From the above, it follows:

$$
\begin{aligned}
|z(t)-u(t)| \leq & \left| z(t) - \tfrac{u_0}{\Gamma(\gamma)} t^{\gamma-1} - \tfrac{\lambda}{\Gamma(\alpha)} \int_0^t (t-s)^{\alpha-1} \frac{f(z(s))}{\left(\int_0^T f(z(x))dx\right)^2} ds \right| \\
& + \tfrac{\lambda}{\Gamma(\alpha)} \int_0^t (t-s)^{\alpha-1} \left| \frac{f(z(s))}{\left(\int_0^T f(z(x))dx\right)^2} - \frac{f(u(s))}{\left(\int_0^T f(u(x))dx\right)^2} \right| ds \\
\leq & \, \epsilon \lambda_\varphi \varphi(t) + \tfrac{\lambda}{\Gamma(\alpha)} \int_0^t (t-s)^{\alpha-1} \frac{1}{\left(\int_0^T f(z(x))dx\right)^2} |f(z(s)) - f(u(s))| \, ds \\
& + \tfrac{\lambda}{\Gamma(\alpha)} \int_0^t (t-s)^{\alpha-1} |f(u(s))| \left| \frac{1}{\left(\int_0^T f(z(x))dx\right)^2} - \frac{1}{\left(\int_0^T f(u(x))dx\right)^2} \right| ds.
\end{aligned} \tag{19}
$$

To substitute (16) and (17) into (19), we get

$$
\begin{aligned}
|z(t)-u(t)| \leq & \, \epsilon \lambda_\varphi \varphi(t) + \frac{\lambda L}{(c_1 T)^2 \Gamma(\alpha)} \int_0^t (t-s)^{\alpha-1} |z(s)-u(s)| \, ds \\
& + \frac{2\lambda c_2^2 T^2 L}{(c_1 T)^4 \Gamma(\alpha)} \int_0^t (t-s)^{\alpha-1} |z(s)-u(s)| \, ds \\
\leq & \, \epsilon \lambda_\varphi \varphi(t) + \left(\frac{\lambda L}{(c_1 T)^2} + \frac{2\lambda c_2^2 T^2 L}{(c_1 T)^4} \right) \frac{1}{\Gamma(\alpha)} \int_0^t (t-s)^{\alpha-1} |z(s)-u(s)| \, ds,
\end{aligned}
$$

and, to apply Lemma 2, we have

$$
|z(t)-u(t)| \leq \left[\left(1 + \nu_1 \lambda_\varphi \left(\frac{\lambda L}{(c_1 T)^2} + \frac{2\lambda c_2^2 T^2 L}{(c_1 T)^4} \right) \right) \lambda_\varphi \right] \epsilon \varphi(t) = C_f \epsilon \varphi(t),
$$

where $\nu_1 = \nu_1(\alpha)$ is a constant. It completes the proof of Theorem 4. □

Acknowledgments: The authors are grateful to anonymous referees for several comments and suggestions.

Author Contributions: All of the authors contributed to the conception and development of this manuscript.

Conflicts of Interest: The authors declare no conflict of interest.

References

1. Kilbas, A.A.; Srivastava, H.M.; Trujillo, J.J. *Theory and Applications of Fractional Differential Equations*; Elsevier: Amsterdam, The Netherlands, 2006; Volume 204.
2. Podlubny, I. *Fractional Differential Equations*; Academy Press: Cambridge, MA, USA, 1999; Volume 198.
3. Benchohra, M.; Henderson, J.; Ntouyas, S.K.; Ouahab, A. Existence results for fractional order functional differential equations with infinite delay. *J. Math. Anal. Appl.* **2008**, *338*, 1340–1350.
4. Benchohra, M.; Bouriah, S. Existence and stability results for nonlinear boundary value problem for implicit differential equations of fractional order. *Moroc. J. Pure Appl. Anal.* **2005**, *1*, 22–37.
5. Ibrahim, R.W. Generalized Ulam–Hyers stability for fractional differential equations. *Int. J. Math.* **2012**, *23*, 1250056; doi:10.1142/S0129167X12500565.
6. Hilfer, R.; Luchko, Y.; Tomovski, Z. Operational method for the solution of fractional differential equations with generalized Riemann-Lioville fractional derivative. *Fract. Calc. Appl. Anal.* **2009**, *12*, 289–318.
7. Vivek, D.; Kanagarajan, K.; Sivasundaram, S. Dynamics and stability of pantograph equations via Hilfer fractional derivative. *Nonlinear Stud.* **2016**, *23*, 685–698.
8. Wang, J.; Lv, L.; Zhou, Y. Ulam stability and data dependence for fractional differential equations with Caputo derivative. *Electron. J. Qual. Theory Differ. Equ.* **2011**, *63*, 1–10.
9. Wang, J.; Zhou, Y. New concepts and results in stability of fractional differential equations. *Commun. Nonlinear Sci. Numer. Simul.* **2012**, *17*, 2530–2538.
10. Hilfer, R. (Ed.) Fractional time evolution. In *Application of Fractional Calculus in Physics*; World Scientific: Singapore, 1999; pp. 87–130.

11. Abbas, S.; Benchohra, M.; Lagreg, J.E.; Alsaedi, A.; Zhou, Y. Existence and Ulam stability for fractional differential equations of Hilfer-Hadamard type. *Adv. Differ. Equ.* **2017**, doi:10.1186/s13662-017-1231-1.
12. Furati, K.M.; Kassim, M.D.; Tatar, N.E. Existence and uniqueness for a problem involving Hilfer fractional derivative. *Comput. Math. Appl.* **2012**, *64*, 1616–1626.
13. Furati, K.M.; Kassim, M.D.; Tatar, N.E. Non-existence of global solutions for a differential equation involving Hilfer fractional derivative. *Electron. J. Differ. Equ.* **2013**, *2013*, 235.
14. Gu, H.; Trujillo, J.J. Existence of mild solution for evolution equation with Hilfer fractional derivative. *Appl. Math. Comput.* **2014**, *257*, 344–354.
15. Wang, J.R.; Yuruo Zhang, Y. Nonlocal initial value problems for differential equations with Hilfer fractional derivative. *Appl. Math. Comput.* **2015**, *266*, 850–859.
16. Samko, S.G.; Kilbas, A.A.; Marichev, O.I. *Fractional Integrals and Derivatives, Theory and Applications*; Gordon and Breach: Amsterdam, The Netherlands, 1987.
17. Kwok, K. *Complete Guide to Semiconductor Devices*; Mc Graw-Hill: New York, NY, USA, 1995.
18. Maclen, E.D. *Thermistors*; Electrochemical Publication: Glasgow, UK, 1979.
19. Khan, A.; Li, Y.; Shah, K.; Khan, T.S. On Coupled p-Laplacian Fractional Differential Equations with Nonlinear Boundary Conditions. *Complexity* **2017**, doi:10.1155/2017/8197610.
20. Song, W.; Gao, W. Existence of solutions for nonlocal p-Laplacian thermistor problems on time scales. *Bound. Value Probl.* **2013**, doi:10.1186/1687-2770-2013-1.
21. Sidi Ammi, M.R.; Torres, D.F.M. Galerkin spectral method for the fractional nonlocal thermistor problem. *Comput. Math. Appl.* **2017**, *73*, 1077–1086.
22. Sidi Ammi, M.R.; Torres, D.F.M. Existence and uniqueness of a positive solution to generalized nonlocal thermistor problems with fractional-order derivatives. *Differ. Equ. Appl.* **2012**, *4*, 267–276.
23. Ammi, M.R.S.; Torres, D.F.M. Existence and uniqueness results for a fractional Riemann-liouville nonlocal thermistor problem on arbitrary time scales. *J. King Saud Univ. Sci.* **2017**, in press.
24. Sidi Ammi, M.R.; Torres, D.F.M. Numerical analysis of a nonlocal parabolic problem resulting from thermistor problem. *Math. Comput. Simul.* **2008**, *77*, 291–300.
25. Sidi Ammi, M.R.; Torres, D.F.M. Optimal control of nonlocal thermistor equations. *Int. J. Control* **2012**, *85*, 1789–1801.
26. Liang, Y.; Chen, W.; Akpa, B.S.; Neuberger, T.; Webb, A.G.; Magin, R.L. Using spectral and cumulative spectral entropy to classify anomalous diffusion in Sephadex^{TM} gels. *Comput. Math. Appl.* **2017**, *73*, 765–774.
27. Andras, S.; Kolumban, J.J. On the Ulam–Hyers stability of first order differential systems with nonlocal initial conditions. *Nonlinear Anal. Theory Methods Appl.* **2013**, *82*, 1–11.
28. Jung, S.M. Hyers-Ulam stability of linear differential equations of first order. *Appl. Math. Lett.* **2004**, *17*, 1135–1140.
29. Muniyappan, P.; Rajan, S. Hyers-Ulam-Rassias stability of fractional differential equation. *Int. J. Pure Appl. Math.* **2015**, *102*, 631–642.
30. Rus, I.A. Ulam stabilities of ordinary differential equations in a Banach space. *Carpathian J. Math.* **2010**, *26*, 103–107.
31. Abbas, S.; Benchohra, M.; Sivasundaram, S. Dynamics and Ulam stability for Hilfer type fractional differential equations. *Nonlinear Stud.* **2016**, *4*, 627–637.
32. Kamocki, R.; Obcznnski, C. On fractional Cauchy-type problems containing Hilfer derivative. *Electron. J. Qual. Theory Differ. Equ.* **2016**, *50*, 1–12.
33. Granas, A.; Dugundji, J. *Fixed Point Theory*; Springer: New York, NY, USA, 2003.
34. Vivek, D.; Kanagarajan, K.; Sivasundaram, S. Theory and analysis of nonlinear neutral pantograph equations via Hilfer fractional derivative. *Nonlinear Stud.* **2017**, *24*, 699–712.

© 2017 by the authors. Licensee MDPI, Basel, Switzerland. This article is an open access article distributed under the terms and conditions of the Creative Commons Attribution (CC BY) license (http://creativecommons.org/licenses/by/4.0/).

fractal and fractional

MDPI

Article

Modeling of Heat Distribution in Porous Aluminum Using Fractional Differential Equation

Rafał Brociek [1,*], Damian Słota [1], Mariusz Król [2], Grzegorz Matula [2] and Waldemar Kwaśny [2]

1 Institute of Mathematics, Silesian University of Technology, Kaszubska 23, 44-100 Gliwice, Poland; damian.slota@polsl.pl

2 Institute of Engineering Materials and Biomaterials, Silesian University of Technology, Konarskiego 18A, 44-100 Gliwice, Poland; mariusz.krol@polsl.pl (M.K.); grzegorz.matula@polsl.pl (G.M.); waldemar.kwasny@polsl.pl (W.K.)

* Correspondence: rafal.brociek@polsl.pl

Received: 20 November 2017; Accepted: 9 December 2017; Published: 12 December 2017

Abstract: The authors present a model of heat conduction using the Caputo fractional derivative with respect to time. The presented model was used to reconstruct the thermal conductivity coefficient, heat transfer coefficient, initial condition and order of fractional derivative in the fractional heat conduction inverse problem. Additional information for the inverse problem was the temperature measurements obtained from porous aluminum. In this paper, the authors used a finite difference method to solve direct problems and the Real Ant Colony Optimization algorithm to find a minimum of certain functional (solve the inverse problem). Finally, the authors present the temperature values computed from the model and compare them with the measured data from real objects.

Keywords: fractional derivative; inverse problem; heat conduction in porous media; thermal conductivity; heat transfer coefficient

---

1. Introduction

Fractional calculus is a part of mathematical analysis and has a lot of applications in technical science. One of the most popular books about fractional calculus is Reference [1]. References [2–4] provide information about fractional calculus, fractional differential equations, approximations of fractional derivatives and numerical methods. There are also a lot of articles about fractional calculus—for example, [5–7].

Various phenomena in nature can be modeled using fractional derivatives [8–16]—for example, in [9,11], the authors surveyed fractional-order electric circuit models, Reference [12] shows applications of fractional derivatives in control theory, and, in [13,14,16,17], we can find information about application fractional derivatives in heat conduction problems. In [16], the authors present an algorithm to solve the fractional heat conduction equation. In the presented model, the heat transfer coefficient is reconstructed based on measurements of the temperature. The direct problem is solved by using the implicit finite difference method. To minimize the functional defining the error of the approximate solution, the Nelder–Mead algorithm is used. In [18], the authors consider the inverse problem of recovering a time-dependent factor of an unknown source on some sub-boundary for a diffusion equation with time fractional derivative. The authors present two regularizing schemes in order to reconstruct an unknown boundary source. Another paper where authors solved the inverse problem with fractional derivatives is Reference [19]. Zhuang et al. considered a time-fractional heat conduction inverse problem with a Caputo derivative in a three-layer composite medium. To solve the direct problem, they used a finite difference method, and, for the inverse problem, the Levenberg–Marquardt method was applied. The results show that the time-fractional heat conduction model provides an effective and accurate simulation of the experimental data. In addition, in [20], it is considered an inverse fractional heat conduction problem. The authors show that the model with fractional derivatives better describes the process of heat conduction in ceramics media.

In this paper, the authors consider the heat conduction inverse problem. A mathematical model describing the heat transfer phenomenon in porous aluminum is given by fractional differential equation with initial-boundary conditions. In this case, we used the Caputo fractional derivative. The algorithm consists of two parts: solution of direct problem and solution of inverse problem by finding the minimum of the functional. Additional information for the inverse problem was the temperature measurements obtained from porous aluminum. The direct problem was solved using a finite difference method and approximations of Caputo derivatives [17,21]. In the inverse problem, the heat transfer coefficient, thermal conductivity coefficient, initial condition and order of derivative were sought. In order to do that, we need to minimize the functional describing the error of approximate solution. The functional was minimized by a Real Ant Colony Optimization algorithm [22,23].

More about heat conduction inverse problems can be found in [24–28]. Zielonka et al. in [25] solved the one-phase inverse problem of alloy solidifying within the casting mould. The authors also include their paper shrinkage of the metal phenomenon.

The investigated inverse problem consists of reconstruction of the heat transfer coefficient on the boundary of the casting mould on the basis of measurements of temperature read from the sensor placed in the middle of the mould. In [28], Stefan problems relevant to burning oil-water systems are considered. The author used the heat balance integral method.

2. Fractional Heat Conduction Equation

In this section, we would like to present the fractional differential equation

$$c\varrho\frac{\partial^\alpha u(x,t)}{\partial t^\alpha} = \lambda\frac{\partial^2 u(x,t)}{\partial x^2}, \tag{1}$$

defined in region $D = \{(x,t) : x \in [0,L], t \in [0,T],\ L,T \in \mathbb{R}_+\}$, where c is specific heat, ϱ is density of material, λ is thermal conductivity coefficient, α is order of derivative and u is the function describing the distribution of temperature. In literature, this equation is called the Time Fractional Diffusion Equation (TFDE). For a more precise description of the model, we still need initial-boundary conditions. In this case, we used a Neumann boundary condition for $x = 0$, and a Robin boundary condition for $x = L$. Below, we present initial-boundary conditions:

$$u(x,0) = f(x), \quad x \in [0,L], \tag{2}$$

$$-\lambda\frac{\partial u}{\partial x}(0,t) = q(t), \quad t \in [0,T], \tag{3}$$

$$-\lambda\frac{\partial u}{\partial x}(L,t) = h(t)(u(L,t) - u^\infty), \quad t \in [0,T], \tag{4}$$

where f is function describing initial condition, q is the heat flux, h is heat transfer coefficient and u^∞ is ambient temperature. By solving models (1)–(4), we obtain the temperature values at the points of the domain D. To model process of heat conduction in porous media, we used Caputo fractional derivative of order $\alpha \in (0,1)$, which, in our case, is defined by the formula [1]:

$$\frac{\partial^\alpha u(x,t)}{\partial t^\alpha} = \frac{1}{\Gamma(1-\alpha)}\int_0^t \frac{\partial u(x,s)}{\partial s}(t-s)^{-\alpha}ds. \tag{5}$$

In the next part of the article, we use the considered model to solve the fractional heat conduction inverse problem.

3. Formulation of the Problem

In an inverse problem, some information in the considered model is unknown; in our case, we do not know: thermal conductivity coefficient λ, heat transfer coefficient h, initial condition f and order of derivative α. We have additional information, called input data, which is the temperature measurements. We denote it by:

$$u(x_p, t_j) = \widehat{U}_j, \ j = 1, 2, \ldots, N_1, \tag{6}$$

where N_1 is number of measurements from thermocouples.

If we solve the direct problems (1)–(4) for fixed values of the sought parameters, then we obtain calculated values of the temperature at certain fixed points of the domain D—in our case, we denote it by $U_j(\lambda, h, f, \alpha)$. Using the calculated values of the temperature $U_j(\lambda, h, f, \alpha)$, input data $\widehat{U}_j$, we create functionals defining the error of approximate solution:

$$F(\lambda, h, f, \alpha) = \sum_{j=1}^{N_1} \left(U_j(\lambda, h, f, \alpha) - \widehat{U}_j\right)^2. \tag{7}$$

By minimizing the functional (7), we find the approximate values of the sought parameters.

The measured temperatures (input data) $\widehat{U}_j$ were obtained from a sample of porous aluminum. This sample was formed by pressurizing the powders' aluminum of medium size 0.8 mm in the plate hydraulic press. Powders were pressed at 150 bar pressure. Sample was heated to 300 °C at speed of 1 K/s and then cooled to ambient temperature. During that time, sample temperature measurements were done, which measurements were used as input data for algorithm.

4. Method of Solution

The solution of the inverse problem can be divided into two parts: first—solution of direct problems (1)–(4), and second—finding minimum of the functional (7).

4.1. Solution of the Direct Problem

In solving the inverse problem, we have to solve the direct problem many times. To solve direct problems (1)–(4), we used implicit finite difference scheme. In order to do that, we create grid

$$S = \left\{(x_i, t_k),\ x_i = i\,\Delta x,\ t_k = k\,\Delta t, i = 0, 1, \ldots, N,\ k = 0, 1, 2, \ldots, M\right\},$$

with size $(N+1) \times (M+1)$ and steps $\Delta x = L/N$, $\Delta t = T/M$. Caputo fractional derivative (5) is approximated by the formula [17]:

$$\frac{\partial^\alpha u}{\partial t^\alpha}(x_i, t_k) \approx D_t^\alpha u_i^k = \sigma(\alpha, \Delta t) \sum_{j=1}^{k} \omega(\alpha, j) \left(u_i^{k-j+1} - u_i^{k-j}\right), \tag{8}$$

where

$$\sigma(\alpha, \Delta t) = \frac{1}{\Gamma(1-\alpha)\,(1-\alpha)\,(\Delta t)^\alpha},$$
$$\omega(\alpha, j) = j^{1-\alpha} - (j-1)^{1-\alpha}.$$

We also need to approximate boundary conditions of the second and third kinds. The following approximations were used:

$$-\lambda_0 \frac{u_1^k - u_{-1}^k}{2\Delta x} = q_k \implies u_{-1}^k = u_1^k + \frac{2\Delta x q_k}{\lambda_0}, \tag{9}$$

$$-\lambda_N \frac{u_{N+1}^k - u_{N-1}^k}{2\Delta x} = h_k (U_N^k - u^\infty) \implies u_{N+1}^k = u_{N-1}^k - \frac{2\Delta x h_k}{\lambda_N}(u_N^k - u^\infty). \tag{10}$$

Using all approximations (8)–(10) and the differential quotient for the derivative of second order with respect to space

$$\frac{\partial^2 u}{\partial x^2}(x_i, t_k) \approx \frac{u_{i-1}^k - 2u_i^k + u_{i+1}^k}{(\Delta x)^2},$$

we get the following differential equations

$k \geq 1,\ i = 0$:

$$\left(\sigma(\alpha,\Delta t) + \frac{2\,a_0}{(\Delta x)^2}\right) u_0^k - \frac{2\,a_0}{(\Delta x)^2}\, u_1^k = \sigma(\alpha,\Delta t)\, u_0^{k-1} - \sigma(\alpha,\Delta t) \sum_{j=2}^{k} \omega(\alpha,j)\left(u_0^{k-j+1} - u_0^{k-j}\right) + \frac{2\,q_k}{c\,\varrho\,\Delta x},$$

$k \geq 1,\ i = 1, 2, \ldots, N-1$:

$$-\frac{a_i}{(\Delta x)^2}\, u_{i-1}^k + \left(\sigma(\alpha,\Delta t) + \frac{2\,a_i}{(\Delta x)^2}\right) u_i^k - \frac{a_i}{(\Delta x)^2}\, u_{i+1}^k$$
$$= \sigma(\alpha,\Delta t)\, u_i^{k-1} - \sigma(\alpha,\Delta t) \sum_{j=2}^{k} \omega(\alpha,j)\left(u_i^{k-j+1} - u_i^{k-j}\right),$$

$k \geq 1,\ i = N$:

$$-\frac{2\,a_N}{(\Delta x)^2}\, u_{N-1}^k + \left(\sigma(\alpha,\Delta t) + \frac{2\,a_N}{(\Delta x)^2} + \frac{2}{c\,\varrho\,\Delta x}\, h_k\right) u_N^k$$
$$= \sigma(\alpha,\Delta t)\, u_N^{k-1} - \sigma(\alpha,\Delta t) \sum_{j=2}^{k} \omega(\alpha,j)\left(u_N^{k-j+1} - u_N^{k-j}\right) + \frac{2}{c\,\varrho\,\Delta x}\, h_k\, u^\infty,$$

where $u_i^k \approx u(x_i, t_k)$, $h_k = h(t_k)$, $q_k = q(t_k)$ and $a_i = \frac{\lambda(x_i)}{c\,\varrho}$ is the thermal diffusivity coefficient. Solving the system of equations gives us approximate values of function u in points of grid S. More about stability of the presented method can be found in [17].

4.2. Minimum of the Functional

The second part of the presented algorithms is finding the minimum of the functional (7). In this paper, we used the Real Ant Colony Optimization algorithm (Algorithm 1). The algorithm was inspired by the behavior of swarm of ants in nature. Pheromone spots, which are L, are identified with solutions. Firstly, they are distributed randomly in the considered area. Then, we rank them according to their quality—better solution means stronger pheromone spot and greater probability to choose it by ant. In this way, we create the solution archive. In every iteration, the one of M ants constructs one new solution (new pheromone spot) using the probability density function (in this case, Gaussian function). The ant chooses with the probability a pheromone spot (solution) and transforms it by sampling its neighborhood using the Gaussian function. Then, the solutions archive is updated with new solutions and sorted according to the quality; next, M worst solutions are rejected. The described algorithm was adapted for parallel computing. For the description of the algorithm, we will introduce the symbols:

$F(\mathbf{x})$	minimized function, $\mathbf{x} = (x_1, \ldots, x_n) \in D$
n	dimension (number of variables)
nT	number of threads
$M = nT \cdot p$	number of ants in population
I	number of iterations
L	number of pheromone spots
q, ξ	parameters of the algorithm

Algorithm 1: Parallel Real ACO algorithm

Initialization of the algorithm

1. Setting input parameters of the algorithm L, M, I, nT, q, ξ.
2. Randomly generate L pheromone spots (solutions) and assign them to set T_0 (starting archive).
3. Calculate values of the minimized function F for each pheromone spot and sort the archive T_0 from best to worst solution.

Iterative process

4. Assigning probabilities to pheromone spots (solutions) according to the following formula:

$$p_l = \frac{\omega_l}{\sum_{l=1}^{L} \omega_l} \quad l = 1, 2, \ldots, \tag{11}$$

where weights ω_l are associated with l-th solution and expressed by the formula

$$\omega_l = \frac{1}{qL\sqrt{2\pi}} \cdot e^{\frac{-(l-1)^2}{2q^2L^2}}.$$

5. Ant chooses a random l-th solution with probability p_l.
6. Ant transforms the j-th coordinate ($j = 1, 2, \ldots, n$) of l-th solution s_j^l by sampling proximity with the probability density function (Gaussian function)

$$g(x, \mu, \sigma) = \frac{1}{\sigma\sqrt{2\pi}} \cdot e^{\frac{-(x-\mu)^2}{2\sigma^2}},$$

where $\mu = s_j^l$, $\sigma = \frac{\xi}{L-1} \sum_{p=1}^{L} |s_j^p - s_j^l|$.
7. Repeat steps 5–6 for each ant. We obtain M new solutions (pheromone spots).
8. Divide new solutions on nT groups. Calculate values of minimized function F for each new solution (parallel computing).
9. Add to the archive T_i new solutions, sort the archive by quality of solutions, remove M worst solution.
10. Repeat steps 4–9 I times.

5. Results

In this section, we present the obtained results. We consider models (1)–(4) with the following data:

$$t \in [0, 71.82]\,[\mathrm{s}],\ x \in [0, 3.825]\,[\mathrm{mm}],\ c = 900 \left[\frac{J}{\mathrm{kg} \cdot \mathrm{K}}\right], \varrho = 2106 \left[\frac{\mathrm{kg}}{\mathrm{m}^3}\right],$$

$$q(t) = 0 \left[\frac{\mathrm{W}}{\mathrm{m}^2}\right],\ u^{\infty} = 298\ [\mathrm{K}],\ x_p = 3.825\,[\mathrm{mm}].$$

In the presented model, we lack the following data:

- $\lambda = a_1 \left[\frac{J}{s^{\alpha} \cdot \mathrm{m} \cdot \mathrm{K}}\right]$—modified thermal conductivity coefficient,
- $f(x) = a_2$ [K]—initial condition,
- $h(t) = a_3 t^2 + a_4 t + a_5 \left[\frac{\mathrm{W}}{\mathrm{m}^2 \cdot \mathrm{K}}\right]$—heat transfer coefficient,
- $\alpha = a_6$—order of derivative,

which is reconstructed based on measurements of temperature. We assume that

$$a_1 \in [100, 300],\ a_2 \in [350, 650],\ a_3 \in [-10, 10],\ a_4 \in [-5, 5],\ a_5 \in [70, 200],\ a_6 \in [0.01, 0.99].$$

Solving the direct problem, we used two different grids 100 × 1995 ($\Delta x = 0.03825$, $\Delta t = 0.036$) and 100 × 3990 ($\Delta x = 0.03825$, $\Delta t = 0.018$). In the case of the real ACO algorithm, we used the following parameters:

$$L = 12, \quad M = 16, \quad I = 55, \quad nT = 4, \quad q = 0.9, \quad \xi = 1.$$

The real ACO algorithm is probabilistic, so we decide to execute them ten times to check the stability.

Table 1 presents values of reconstructed parameters $a_1, a_2, \ldots, a_6$ in the case of two different grids. In both grids, parameters have similar values except parameter a_1—the thermal conductivity coefficient. For 100 × 1995 grid, a_1 is equal to 300 and, for 100 × 3990 grid, the parameter has value 237.91. The initial condition is approximately equal to 569 K and the order of derivative is equal to 0.20 (100 × 1995) or 0.21 (100 × 3990). As we can see in Table 1, if the 100 × 1995 grid is used, then, for the most of sought after parameters, the standard deviation is less than for the 100 × 3990 grid. This means that, for the first grid, the obtained results from the ACO algorithm were slightly less dispersed than in the case of the second grid.

Table 1. Results of calculation for grids 100 × 1995 and 100 × 3990 (a_i—reconstructed value of parameter, σ_{a_i}—standard deviation ($i = 1, 2, \ldots, 6$)).

	100×1995	σ_{a_i}	**100×3990**	σ_{a_i}
a_1	300.00	69.74	237.91	67.78
a_2	569.73	2.02	566.74	3.80
a_3	1.63	0.40	1.52	2.20
a_4	4.72	0.67	5.00	4.27
a_5	198.02	46.05	178.05	51.73
a_6	0.20	0.05	0.21	0.09
value of the functional	246.98		352.88	

In Figure 1, we can see plots of reconstructed function h for two grids. Both plots are similar. The values of the function for the grid 100 × 1995 are slightly larger than for the second grid. Table 2 presents errors of reconstruction temperature in measurement points. In both cases, these errors are similar and temperature is reconstructed very well. Average errors are a little bit smaller in the case of 100 × 3990 grid, but, in the case of maximal errors, it is otherwise.

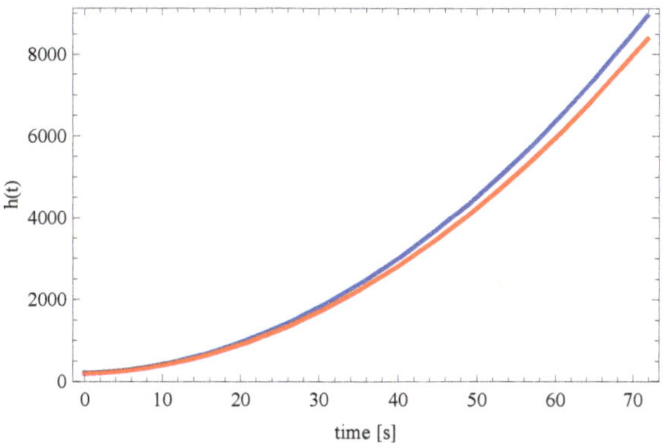

Figure 1. Plots of reconstructed function h for grid 100 × 1995 (blue line) and grid 100 × 3990 (red line).

Table 2. Errors of temperature reconstruction in measurement point $x_p = 3.825$ for grids 100×1995 and 100×3990 (Δ_{avg}—average absolute error, Δ_{max}—maximal absolute error, δ_{avg}—average relative error, δ_{max}—maximal relative error).

	100×1995	**100×3990**
$\Delta_{avg}[K]$	4.92	4.77
$\Delta_{max}[K]$	11.04	12.38
$\delta_{avg}[\%]$	1.06	1.02
$\delta_{max}[\%]$	3.08	3.46

At the end of this section, we would like to present plots of reconstructed temperature and distribution of errors of this reconstruction (Figures 2 and 3). Temperature reconstruction and distribution of errors are very similar.

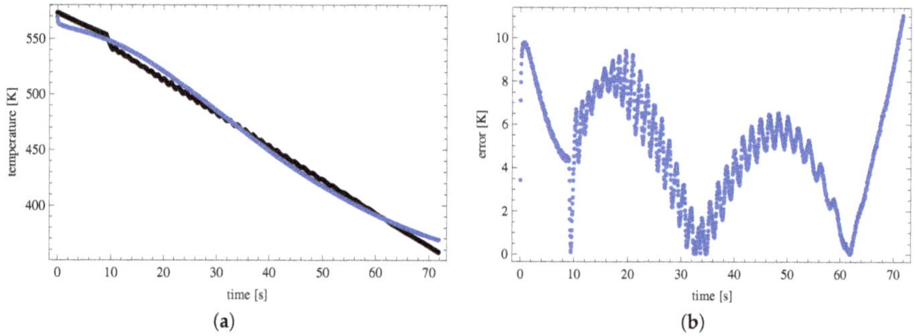

Figure 2. (**a**) Measurements of the temperature (black line) and reconstructed temperature (blue line); and (**b**) distribution of errors for this reconstruction (100 × 1995) .

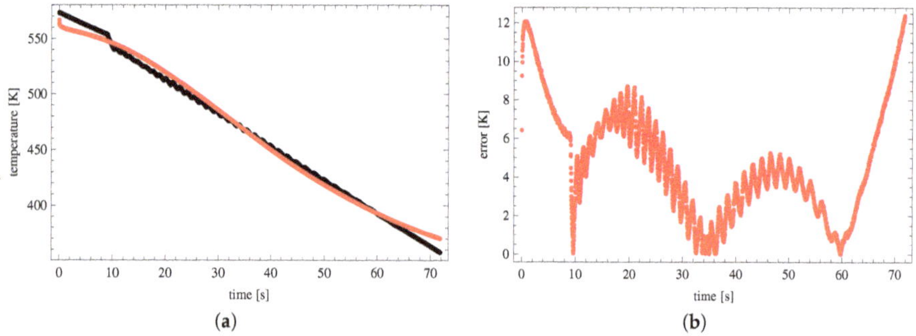

Figure 3. (**a**) Measurements of the temperature (black line), reconstructed temperature (red line); and (**b**) distribution of errors for this reconstruction (100 × 3990).

6. Conclusions

In summary, the paper presents a fractional heat conduction equation with Caputo derivative. Based on this equation, the inverse problem is solved using temperature measurements from porous aluminum. Next, computed values of temperature were compared with real data. First of all, the results obtained were very good. Average relative error of reconstruction was equal to 1.06% for 100 × 1995 grid, and 1.02% for 100 × 3990 grid. More density grid gives us smaller relative errors, but a little bit higher maximal errors. Using less density, grid results obtained from the ACO algorithm were less dispersed. The sought order of fractional derivative is approximately 0.20. In the future, we would like to take under consideration other models of heat conduction using different fractional derivatives.

Acknowledgments: The authors would like to thank the reviewers for their constructive comments and suggestions.

Author Contributions: Mariusz Król, Grzegorz Matula and Waldemar Kwaśny created a sample of porous aluminum, and then made an experiment involving the heating and cooling of the sample. Temperature measurements are the result of experiment; Rafał Brociek and Damian Słota have developed an algorithm for solving direct and inverse problems. They performed a numerical experiment , analyzed the data and wrote the paper.

Conflicts of Interest: The authors declare no conflict of interest.

References

1. Podlubny, I. *Fractional Differential Equations*; Academic Press: San Diego, CA, USA, 1999.
2. Guo, B.; Pu, X.; Huang, F. *Fractional Partial Differential Equations and Their Numerical Solutions*; World Scientific: Singapore, 2015.
3. Ortigueira, M.D. *Fractional Calculus for Scientists and Engineers*; Springer: Berlin, Germany, 2011.
4. Das, S. *Functional Fractional Calculus for System Identification and Controls*; Springer: Berlin, Germany, 2008.
5. Povstenko, Y.; Klekot, J. The Cauchy problem for the time-fractional advection diffusion equation in a layer. *Tech. Sci.* **2016**, *19*, 231–244.
6. Abdeljawat, T. On conformable fractional calculus. *J. Comput. Appl. Math.* **2015**, *279*, 57–66.
7. Liu, H.Y.; He, J.H.; Li, Z.B. Fractional calculus for nanoscale flow and heat transfer. *Int. J. Numer. Methods Heat Fluid Flow* **2014**, *24*, 1227–1250.
8. Atangana, A. On the new fractional derivative and application to nonlinear Fisher's reaction–diffusion equation. *Appl. Math. Comput.* **2016**, *273*, 948–956.
9. Freeborn, T.J.; Maundy, B.; Elwakil, A.S. Fractional-order models of supercapacitors, batteries and fuel cells: A survey. *Mater. Renew. Sustain. Energy* **2015**, *4*, doi:10.1007/s40243-015-0052-y.
10. Błasik, M.; Klimek, M. Numerical solution of the one phase 1D fractional Stefan problem using the front fixing method. *Math. Methods Appl. Sci.* **2014**, *38*, 3214–3228.
11. Mitkowski, W.; Skruch, P. Fractional-order models of the supercapacitors in the form of RC ladder networks. *J. Pol. Acad. Sci.* **2013**, *61*, 581–587.
12. Matušů, R. Application of fractional order calculus to control theory. *Int. J. Math. Models Methods Appl. Sci.* **2011**, *5*, 1162–1169.
13. Povstenko, Y.Z. Fractional heat conduction equation and associated thermal stress. *J. Therm. Stresses* **2004**, *28*, 83–102.
14. Yang, X.J.; Baleanu, D. Fractal heat conduction problem solved by local fractional variation iteration method. *Therm. Sci.* **2013**, *17*, 625–628.
15. Murio, D.A. Time fractional IHCP with Caputo fractional derivatives. *Comput. Math. Appl.* **2008**, *56*, 2371–2381.
16. Brociek, R.; Słota, D. Reconstruction of the boundary condition for the heat conduction equation of fractional order. *Therm. Sci.* **2015**, *19*, 35–42.
17. Murio, D.A. Implicit finite difference approximation for time fractional diffusion equations. *Comput. Math. Appl.* **2008**, *56*, 1138–1145.
18. Liu, J.J.; Yamamoto, M.; Yan, L.L. On the reconstruction of unknown time-dependent boundary sources for time fractional diffusion process by distributing measurement. *Inverse Probl.* **2016**, *32*, doi:10.1088/0266-5611/32/1/015009.
19. Zhuag, Q.; Yu, B.; Jiang, X. An inverse problem of parameter estimation for time-fractional heat conduction in a composite medium using carbon–carbon experimental data. *Physica B* **2015**, *456*, 9–15.
20. Obrączka, A.; Kowalski, J. Modelowanie rozkładu ciepła w materiałach ceramicznych przy użyciu równań różniczkowych niecałkowitego rzędu. In Proceedings of the Materiały XV Jubileuszowego Sympozjum "Podstawowe Problemy Energoelektroniki, Elektromechaniki i Mechatroniki", Gliwice, Poland, 11–13 December 2012; Volume 32. (In Polish)
21. Brociek, R. Implicite finite difference metod for time fractional diffusion equations with mixed boundary conditions. *Zesz. Naukowe Politech. Śląskiej* **2014**, *4*, 73–87. (In Polish)

22. Brociek, R.; Słota, D. Application of real ant colony optimization algorithm to solve space and time fractional heat conduction inverse problem. *Inf. Technol. Control* **2017**, *46*, 5–16.
23. Socha, K.; Dorigo, M. Ant Colony Optimization in continuous domains. *Eur. J. Oper. Res.* **2008**, *185*, 1155–1173.
24. Hetmaniok, E.; Hristov, J.; Słota, D.; Zielonka, A. Identification of the heat transfer coefficient in the two-dimensional model of binary alloy solidification. *Heat Mass Transf.* **2017**, *53*, 1657–1666.
25. Zielonka, A.; Hetmaniok, E.; Słota, D. Inverse alloy solidification problem including the material shrinkage phenomenon solved by using the bee algorithm. *Int. Commun. Heat Mass Transf.* **2017**, *87*, 295–301.
26. Zhang, B.; Qi, H.; Ren, Y.-T.; Sun, S.-C.; Ruan, L.-M. Application of homogenous continuous Ant Colony Optimization algorithm to inverse problem of one-dimensional coupled radiation and conduction heat transfer. *Int. J. Heat Mass Transf.* **2013**, *66*, 507–516.
27. Grysa, K.; Leśniewska, R. Different finite element approaches for inverse heat conduction problems. *Inverse Probl. Sci. Eng.* **2010**, *18*, 3–17.
28. Hristov, J. An inverse Stefan problem relevant to boilover: Heat balance integral solutions and analysis. *Therm. Sci.* **2007**, *11*, 141–160.

© 2017 by the authors. Licensee MDPI, Basel, Switzerland. This article is an open access article distributed under the terms and conditions of the Creative Commons Attribution (CC BY) license (http://creativecommons.org/licenses/by/4.0/).

fractal and fractional

MDPI

Article

Monitoring Liquid-Liquid Mixtures Using Fractional Calculus and Image Analysis

Ervin K. Lenzi [1], Andrea Ryba [2] and Marcelo K. Lenzi [3,*]

1 Departamento de Física, Universidade Estadual de Ponta Grossa, Av. Carlos Cavalcanti, 4748, Ponta Grossa 84030-900, Paraná, Brazil; eklenzi@uepg.br
2 Departamento de Transportes, Universidade Federal do Paraná, Rua Coronel Francisco H. dos Santos, 100, Curitiba 81531-980, Parana, Brazil; andrea.ryba@ufpr.br
3 Departamento de Engenharia Química, Universidade Federal do Paraná, Rua Coronel Francisco H. dos Santos, 100, Curitiba 81531-980, Parana, Brazil
* Correspondence: lenzi@ufpr.br; Tel.: +55-41-3361-3577

Received: 28 December 2017; Accepted: 8 February 2018; Published: 11 February 2018

Abstract: A fractional-calculus-based model is used to analyze the data obtained from the image analysis of mixtures of olive and soybean oil, which were quantified with the RGB color system. The model consists in a linear fractional differential equation, containing one fractional derivative of order α and an additional term multiplied by a parameter k. Using a hybrid parameter estimation scheme (genetic algorithm and a simplex-based algorithm), the model parameters were estimated as $k = 3.42 \pm 0.12$ and $\alpha = 1.196 \pm 0.027$, while a correlation coefficient value of 0.997 was obtained. For the sake of comparison, parameter α was set equal to 1 and an integer order model was also studied, resulting in a one-parameter model with $k = 3.11 \pm 0.28$. Joint confidence regions are calculated for the fractional order model, showing that the derivative order is statistically different from 1. Finally, an independent validation sample of color component B equal to 96 obtained from a sample with olive oil mass fraction equal to 0.25 is used for prediction purposes. The fractional model predicted the color B value equal to 93.1 ± 6.6.

Keywords: liquid-liquid mixture; fractional calculus; parameter estimation; image analysis; RGB; olive oil

1. Introduction

Process monitoring represents an important and fundamental tool aimed at process safety and economics while meeting environmental regulations. However, for suitable process monitoring, the analytical instrumentation remains a challenge due to higher costs, equipment sensitivity, and level of detection, among others [1]. In order to overcome some of these difficulties, image analysis represents an important field of process monitoring [2]. The main advantages and features are the low cost, non-invasive characteristics, and high precision/accuracy [3]. The literature reports a broad range of image analysis applications to food science [4], medical science [5], road-traffic monitoring [6], road pavement monitoring [7], flare combustion [8], and composition monitoring [9]. Additionally, applications in the quantification of edible oils mixtures, which have applications in many processes, such as biodiesel synthesis [10], pavement rejuvenating agents [11,12], and polyurethane synthesis [13], has also been reported. In this sense, Fernandes et al. [14] analyzed olive and soybean oil mixtures using image analysis (RGB color system) with linear models for a range of olive oil mass fraction of 0–0.7. When coupling the image analysis to Ultraviolet–visible spectroscopy (UV–VIS) spectra, successful predictions in the range of olive oil mass fraction of 0–1 could be obtained. Therefore, the use of nonlinear models would avoid the use of the UV–VIS spectra in order to accurately predict the mixture content, as the values of one of the color components behaved as an exponential decay with an increase in the olive oil content. This would be an important feature as experimental steps would vanish, therefore, leading to faster and cheaper results.

Towards this, a mathematical model that consists of a linear fractional differential equation is proposed. It contains one fractional derivative of order α and one other linear term multiplied by a parameter k. It was used to analyze the Fernandes et al. [14] experimental data. It is worth mentioning that fractional calculus deals with differential operators of arbitrary order, being an important tool to describe memory and hereditary effects of properties of many materials and processes [15]. Fractional calculus also presents a broad range of applications in process systems engineering, rheology, viscoelasticity, acoustics, optics, chemical physics, robotics, electrical engineering, bioengineering, anomalous diffusion [16–21]. The solution of the model here proposed is a nonlinear algebraic equation based on the Mittag-Leffer function, which according to the value of parameter α can turn into, for example, an exponential function. We also performed a comparison with an integer order model in order to show that the fractional derivative order is more suitable to describe the experimental data behavior.

2. Materials and Methods

The experimental data previously reported by Fernandes et al. in [14] concerning the off-line monitoring of mixtures of olive and soybean oil was used in this manuscript. To summarize, Fernandes et al. [14] used the RGB color system for off-line image analysis, where a given number of samples was used for parameter estimation of linear models aimed at concentration prediction and an independent sample was used for validation purposes, further details can be found in reference [14]. Below, the Table 1 presents the experimental data used in this work.

Table 1. Experimental data.

Sample Purpose	Olive Oil Mass Fraction	Color Component B
	0	175
	0.1	144
	0.2	116
	0.3	83
	0.4	54
Parameter Estimation	0.5	29
	0.6	12
	0.7	0.8
	0.8	0
	0.9	0
	1	0
Model Validation	0.25	96

The main idea of the model is to consider the color component B as a function of the mass fraction (mf) of the olive oil in the mixture, as if it would be "disappearing" with an increase of the olive oil content. Secondly, it is assumed that the color component behavior follows the fractional differential equation of order α, subjected to the initial conditions given by $B(mf = 0) = B_0$ and $\frac{d}{d(mf)}B(mf = 0) = 0$:

$$\frac{d^{\alpha}B(mf)}{d(mf)^{\alpha}} = -(k\,B(mf)), \tag{1}$$

where the fractional derivative operator considered here is the Caputo [22] sense, defined as $\frac{d^{\alpha}f(x)}{dx^{\alpha}} = \frac{1}{\Gamma(n-\alpha)}\int_0^x \frac{f^{(n)}(\tau)}{(x-\tau)^{\alpha-n+1}}\,d\tau$, with $f^{(n)}(x) = \frac{d^n f(x)}{dx^n}$ and $(n-1) < \alpha < n$, where n is an arbitrary integer number, and α is a real number.

The solution of Equation (1) can be obtained by using the Laplace Transform Method and it can be expressed in terms of the Mittag-Leffer function [23] as follows:

$$B(mf) = B_0\, E_{\alpha,\alpha}\left(-k\,(mf)^{\alpha}\right) = B_0\left[\sum_{j=0}^{\infty}\frac{\left(-k\,(mf)^{\alpha}\right)^{j}}{\Gamma(\alpha\,(j+1))}\right] = B_0\left[\sum_{j=0}^{\infty}\frac{(-1)^{j}\,(k)^{j}\,(mf)^{\alpha\times j}}{\Gamma(\alpha\,(j+1))}\right], \quad (2)$$

where Γ is the Gamma function (see Equation (17)). It is important to observe that for $\alpha = 1$, Equation (2) turns into an integer order model (for $B(mf = 0) = B_0$), which basically concerns an exponential variation of the color component, i.e.:

$$\frac{dB(mf)}{d(mf)} = -(k\,B(mf)) \rightarrow B(mf) = B_0\, e^{-[k\,(mf)]}. \quad (3)$$

The parameter estimation procedure considers a hybrid task aimed at minimizing the least square function, given by:

$$F_{\text{OBJ}} = \sum_{i=1}^{\text{NE}}(delta_i(mf))^2 = \sum_{i=1}^{\text{NE}}\left(B_i^{\text{EXP}}(mf) - B_i^{\text{MOD}}(mf)\right)^2$$
$$F_{\text{OBJ}} = \sum_{i=1}^{\text{NE}}\left(B_i^{\text{EXP}}(mf) - \left(B_0\left[\sum_{j=0}^{\infty}\frac{(-1)^{j}\,(k)^{j}\,(mf)^{\alpha\times j}}{\Gamma(\alpha\,(j+1))}\right]\right)\right)^2 \quad (4)$$

where NE is the number of experimental data, EXP refers to experimental data and MOD refers to model predictions and delta is the difference between experimental data and model predictions.

A genetic algorithm, based on Isfer et al. [24], was firstly used and the results of this initial estimation step were used as the initial guess of a simplex-based method [25], in order to refine the solution. The main role of the genetic algorithm is to avoid a local minimum of the objective function. Regarding the integer order model, the parameter estimation task used the following expression as an objective function:

$$F_{\text{OBJ}} = \sum_{i=1}^{\text{NE}}\left(B_i^{\text{EXP}}(mf) - \left(B_0\, e^{-(k\,(mf))}\right)\right)^2. \quad (5)$$

The statistical analysis of the estimated parameters and the mathematical model itself followed previously-reported procedures [26–28]. The parametric variance matrix, $\underline{\underline{\mathbf{V}_{\text{param}}}}$, is defined as for the fractional order model, where parameters α and k need to be estimated. From this matrix one can obtain the standard deviation of the model parameters:

$$\left[\underline{\underline{\mathbf{V}_{\text{param}}}}\right]_{(2\times2)} = \begin{bmatrix}\sigma_{\text{k}}^2 & \sigma_{\text{k}-\alpha}^2\\ \sigma_{\text{k}-\alpha}^2 & \sigma_{\alpha}^2\end{bmatrix}_{(2\times2)}. \quad (6)$$

Regarding the integer order model, the parameter variance is given by:

$$\left[\underline{\underline{\mathbf{V}_{\text{param}}}}\right]_{(1\times1)} = \left[\sigma_k^2\right]_{(1\times1)}, \quad (7)$$

where only the parameter k needs to be estimated.

From the parametric variance matrix, one can obtain the parametric correlation matrix, given by:

$$\left[\underline{\underline{\mathbf{r}_{\text{param}}}}\right]_{(2\times2)} == \begin{bmatrix}1 & \frac{\sigma_{k-\alpha}^2}{\sigma_k\sigma_\alpha}\\ \frac{\sigma_{k-\alpha}^2}{\sigma_k\sigma_\alpha} & 1\end{bmatrix}_{(2\times2)}, \quad (8)$$

for the fractional order matrix, while for the integer order matrix, this matrix does not exist as the model has only one parameter. It is important to emphasize that the experimental errors (variance) are considered constant and equal for all experimental data and its value is adequately predicted by:

$$\sigma^2_{B\text{EXP}} = \frac{F_{\text{OBJ}}}{\text{NE} - \text{NP}}, \tag{9}$$

where NE is the number of experiments and NP is the number of parameter to be estimated ("k" and "α" (alpha) in the case of the fractional model and "k" in the case of the integer order model). Note that a better value of the experimental error can be obtained by experimental runs carried out in triplicate or quadruplicate.

Therefore, an experimental variance matrix, $\underline{\underline{\mathbf{V}_{B\text{EXP}}}}$, can be obtained, given by Equation (10), where $\underline{\underline{\mathbf{I}}}$ is a square identity matrix of dimension NE × NE:

$$\left[\underline{\underline{\mathbf{V}_{B\text{EXP}}}}\right]_{(\text{NE}\times\text{NE})} = \begin{bmatrix} \sigma^2_{B\text{EXP}} & 0 & \cdots & 0 \\ 0 & \sigma^2_{B\text{EXP}} & \cdots & 0 \\ \vdots & \vdots & \ddots & \vdots \\ 0 & 0 & \cdots & \sigma^2_{B\text{EXP}} \end{bmatrix}_{(\text{NE}\times\text{NE})} = \sigma^2_{B\text{EXP}} \left[\underline{\underline{\mathbf{I}}}\right]_{(\text{NE}\times\text{NE})}. \tag{10}$$

According to the literature [24–28], the elements of the matrix given by Equations (6) and (7) are given by Equation (11):

$$\left[\underline{\underline{\mathbf{V}_{\text{param}}}}\right]_{(2\times2)} \cong \sigma^2_{y\text{EXP}} \left[\underline{\underline{\mathbf{H}(F_{\text{OBJ}})_{\text{param}}}}\right]^{-1}_{(2\times2)} = \sigma^2_{B\text{EXP}} \begin{pmatrix} \left[\underline{\underline{\mathbf{G}}}\right]^T_{(2\times\text{NE})} \left[\underline{\underline{\mathbf{G}}}\right]_{(\text{NE}\times2)} \\ -2\sum_{i=1}^{\text{NE}} \left[\underline{\underline{\mathbf{H}(B_i^{\text{MOD}})_{\text{param}}}}\right]_{(2\times2)} delta_i(mf) \end{pmatrix}^{-1}, \tag{11}$$

where $\mathbf{H}(F_{\text{OBJ}})_{param}$ is the Hessian matrix of the objective function (Equation (4) or Equation (5)) where the derivatives were obtained with respect to the parameters and $\mathbf{H}(B_i^{\text{MOD}})_{\text{param}}$ is the Hessian matrix of the model (Equation (2) or Equation (3)) where the derivatives were also obtained with respect to the parameters. As the model predictions are usually close to the experimental values, $delta_i(mf)$ is close to zero, the parametric variance matrix is commonly simplified [28] to the following equation:

$$\left[\underline{\underline{\mathbf{V}_{\text{param}}}}\right]_{(2\times2)} \cong \sigma^2_{B\text{EXP}} \left[\underline{\underline{\mathbf{H}(F_{\text{OBJ}})_{\text{param}}}}\right]^{-1}_{(2\times2)} = \sigma^2_{B\text{EXP}} \left(\left[\underline{\underline{\mathbf{G}}}\right]^T_{(2\times\text{NE})} \left[\underline{\underline{\mathbf{G}}}\right]_{(\text{NE}\times2)} \right)^{-1}. \tag{12}$$

The sensitivity matrix $\mathbf{G}$, present in Equation (12), regarding the fractional order model can be written as:

$$\left[\underline{\underline{\mathbf{G}}}\right]_{(\text{NE}\times2)} = \begin{bmatrix} \frac{\partial B_1^{\text{MOD}}}{\partial k} & \frac{\partial B_1^{\text{MOD}}}{\partial \alpha} \\ \frac{\partial B_2^{\text{MOD}}}{\partial k} & \frac{\partial B_2^{\text{MOD}}}{\partial \alpha} \\ \vdots & \vdots \\ \frac{\partial B_{\text{NE}}^{\text{MOD}}}{\partial k} & \frac{\partial B_{\text{NE}}^{\text{MOD}}}{\partial \alpha} \end{bmatrix}_{(\text{NE}\times2)}. \tag{13}$$

Concerning the integer order model, it is given by:

$$\left[\underline{\underline{\mathbf{G}}}\right]_{(\text{NE}\times1)} = \begin{bmatrix} \frac{\partial B_1^{\text{MOD}}}{\partial k} \\ \frac{\partial B_2^{\text{MOD}}}{\partial k} \\ \vdots \\ \frac{\partial B_{\text{NE}}^{\text{MOD}}}{\partial k} \end{bmatrix}_{(\text{NE}\times1)}. \tag{14}$$

Considering the fractional case, the model derivatives c with respect to the parameters are given by the expressions:

$$\frac{\partial B_i^{\mathrm{MOD}}(mf)}{\partial k} = B_0\left[\sum_{j=0}^{\infty}\frac{(-1)^j j(k)^{j-1}(mf)^{\alpha\times j}}{\Gamma(\alpha(j+1))}\right], \tag{15}$$

$$\frac{\partial B_i^{\mathrm{MOD}}(mf)}{\partial \alpha} = B_0\sum_{j=0}^{\infty}\left(\frac{(-1)^j (k)^j (mf)^{\alpha\times j}(j)(\ln(mf)-\Psi(\alpha(j+1)))}{\Gamma(\alpha(j+1))}\right), \tag{16}$$

where the relations defined below are valid [29], where x and t are dummy variables:

$$\begin{gathered}\Gamma(x)=\int_0^{\infty} e^{-t}t^{x-1}dt,\\ \frac{d\Gamma(x)}{dx}=\int_0^{\infty} e^{-t}t^{x-1}\ln(t)dt,\\ \Psi(x)=\frac{1}{\Gamma(x)}\frac{d\Gamma(x)}{dx}.\end{gathered} \tag{17}$$

On the other hand, considering the integer order model, the model derivative with respect to the parameter is given by Equation (18):

$$\frac{\partial B_i^{\mathrm{MOD}}(mf)}{\partial k} = -B_0(mf)e^{-k(mf)}. \tag{18}$$

The parameters confidence intervals are calculated by:

$$\begin{cases} k_{\mathrm{estimated}} - \left[t_{\mathrm{NE-NP}}^{1-\frac{\beta}{2}}(\sigma_k)\right] < k < k_{\mathrm{estimated}} + \left[t_{\mathrm{NE-NP}}^{1-\frac{\beta}{2}}(\sigma_k)\right] \\ \alpha_{\mathrm{estimated}} - \left[t_{\mathrm{NE-NP}}^{1-\frac{\beta}{2}}(\sigma_\alpha)\right] < \alpha < \alpha_{\mathrm{estimated}} + \left[t_{\mathrm{NE-NP}}^{1-\frac{\beta}{2}}(\sigma_\alpha)\right] \end{cases}, \tag{19}$$

for the fractional order model and by:

$$\left\{k_{\mathrm{estimated}} - \left[t_{\mathrm{NE-NP}}^{1-\frac{\beta}{2}}(\sigma_k)\right] < k < k_{\mathrm{estimated}} + \left[t_{\mathrm{NE-NP}}^{1-\frac{\beta}{2}}(\sigma_k)\right]\right., \tag{20}$$

for the integer order model. The parameter standard deviation is obtained from the square root of the elements of the diagonal of the matrix given by Equation (12). Additionally, for a given confidence level (usually 95%, therefore, β equals 0.05) and the degree of freedom (NE–NP), the Student's t distribution value is used to obtain the formal confidence interval.

The parameter joint confidence region of the fractional order model is given by:

$$\begin{bmatrix} k-k_{estimated} & \alpha-\alpha_{\mathrm{estimated}}\end{bmatrix}_{(1\times2)}^{T}\left[\underline{\underline{\mathbf{H}(F_{\mathrm{OBJ}})_{\mathrm{param}}}}\right]_{(2\times2)}\begin{bmatrix} k-k_{\mathrm{estimated}} \\ \alpha-\alpha_{\mathrm{estimated}}\end{bmatrix}_{(2\times1)} = \sigma_{\beta\mathrm{EXP}}^2\,\mathrm{NP}\,F_{(2,\mathrm{NE}-2)}^{1-\beta}, \tag{21}$$

where the value of the Fisher distribution value, F, is obtained for a given confidence level (usually 95%, therefore β equals 0.05) and the degrees of freedom NP and (NE−NP). In the previous equation, the approximation given by Equation (12) is considered, consequently, Equation (22), is used in this work for the fractional order model:

$$\sigma_{\beta\mathrm{EXP}}^2\left[\underline{\underline{\mathbf{H}(F_{\mathrm{OBJ}})_{\mathrm{param}}}}\right]_{(2x2)}^{-1} = \sigma_{\beta\mathrm{EXP}}^2\left(\left[\underline{\underline{\mathbf{G}}}\right]_{(2\times\mathrm{NE})}^{T}\left[\underline{\underline{\mathbf{G}}}\right]_{(\mathrm{NE}\times2)}\right)^{-1} \to \left[\underline{\underline{\mathbf{H}(F_{\mathrm{OBJ}})_{\mathrm{param}}}}\right]_{(2\times2)} = \left(\left[\underline{\underline{\mathbf{G}}}\right]_{(2\times\mathrm{NE})}^{T}\left[\underline{\underline{\mathbf{G}}}\right]_{(\mathrm{NE}\times2)}\right). \tag{22}$$

It is important to emphasize that the region calculated by the equation

$$\begin{bmatrix} k - k_{\text{estimated}} & \alpha - \alpha_{\text{estimated}} \end{bmatrix}^T_{(1\times2)} \left[\left(\left[\underline{\underline{\mathbf{G}}}\right]^T_{(2\times\text{NE})}\left[\underline{\underline{\mathbf{G}}}\right]_{(\text{NE}\times2)}\right)\right]_{(2\times2)} \begin{bmatrix} k - k_{\text{estimated}} \\ \alpha - \alpha_{\text{estimated}} \end{bmatrix}_{(2\times1)} = \sigma^2_{B\text{EXP}}\,\text{NP}\,F^{1-\beta}_{(2,\text{NE}-2)}, \quad (23)$$

usually has an ellipsoidal shape due to the presence of parameter correlation and corresponds to a linearization of the objective function [24–28].

A more realistic parametric joint confidence region can be obtained by Equation (24), where the nonlinear feature of the objective function is taken into account. It is important to stress that the region calculated by Equation (24) is larger than the region obtained by Equation (23) and it does not necessarily has an ellipsoidal shape [30]:

$$F_{\text{OBJ}}(k,\alpha) = F_{\text{OBJ}}(k_{\text{estimated}}, \alpha_{\text{estimated}})\left(1 + \left(\frac{\text{NP}}{\text{NE}-\text{NP}}\right)F^{1-\beta}_{(\text{NP},\text{NE}-\text{NP})}\right), \quad (24)$$

where $F_{\text{OBJ}}(k_{\text{estimated}}, \alpha_{\text{estimated}})$ is the final value of the objective function.

The correlation coefficient between model predictions and experimental data, r, is:

$$r = \frac{\sum_{i=1}^{\text{NE}}\left(B_i^{\text{EXP}}(mf) - \overline{B_i^{\text{EXP}}(mf)}\right)\left(B_i^{\text{MOD}}(mf) - \overline{B_i^{\text{MOD}}(mf)}\right)}{\sqrt{\left(\sum_{i=1}^{\text{NE}}\left(B_i^{\text{EXP}}(mf) - \overline{B_i^{\text{EXP}}(mf)}\right)^2\right)\left(\sum_{i=1}^{\text{NE}}\left(B_i^{\text{MOD}}(t) - \overline{B_i^{\text{MOD}}(mf)}\right)^2\right)}}, \overline{B(mf)} = \sum_{i=1}^{\text{NE}}\frac{B_i(mf)}{\text{NE}}. \quad (25)$$

Finally, regarding the fractional order model, the variance of model predictions of the experimental data used in the parameter estimation task, named in this work as model confidence region, is given by Equations (26) and (27). If the confidence interval of future experimental data prediction is necessary, Equations (28) and (29) should be used instead as, in this case, both the experimental error and the model error prediction should be considered. The following equations were also used for the integer order model analysis:

$$B_i^{\text{EXP}} - t^{1-\frac{\beta}{2}}_{\text{NE}-\text{NP}}\sqrt{V^{\text{MOD}}_{B\text{EXP}}(i,i)} < B_i^{\text{EXP}} < B_i^{\text{EXP}} + t^{1-\frac{\beta}{2}}_{\text{NE}-\text{NP}}\sqrt{V^{\text{MOD}}_{B\text{EXP}}(i,i)}, \quad (26)$$

$$\left[\underline{\underline{\mathbf{V}^{\text{MOD}}_{B\text{EXP}}}}\right]_{(\text{NE}\times\text{NE})} = \left[\underline{\underline{\mathbf{G}}}\right]_{(\text{NE}\times2)}\left[\underline{\underline{\mathbf{V}_{\text{param}}}}\right]_{(2\times2)}\left[\underline{\underline{\mathbf{G}}}\right]^T_{(2\times\text{NE})}, \quad (27)$$

$$B_i^{\text{EXP}} - t^{1-\frac{\beta}{2}}_{\text{NE}-\text{NP}}\sqrt{\mathbf{V}^{\text{PRED}}_{B\text{EXP}}(i,i)} < B_i^{\text{EXP}} < B_i^{\text{EXP}} + t^{1-\frac{\beta}{2}}_{\text{NE}-\text{NP}}\sqrt{\mathbf{V}^{\text{PRED}}_{B\text{EXP}}(i,i)}, \quad (28)$$

$$\left[\underline{\underline{\mathbf{V}^{\text{PRED}}_{B\text{EXP}}}}\right]_{(\text{NE}\times\text{NE})} = \left[\underline{\underline{\mathbf{G}}}\right]_{(\text{NE}\times2)}\left[\underline{\underline{\mathbf{V}_{\text{param}}}}\right]_{(2\times2)}\left[\underline{\underline{\mathbf{G}}}\right]^T_{(2\times\text{NE})} + \left[\underline{\underline{\mathbf{V}_{B\text{EXP}}}}\right]_{(\text{NE}\times\text{NE})}. \quad (29)$$

3. Results

For the genetic algorithm, the number of individuals (each individual consist on a pair of k, α for the fractional order model, or a single value of k for the integer order model) was set equal to 200 and the number of generations was set equal to 100. The probabilities of crossover and mutation were set as 80% assuring a good macroscopic search and of 10% assuring a good microscopic (refinement), respectively. In order to obtain each individual of a given generation, two individuals ("genitors") of the previous generation were randomly selected. After that, a random number in the interval (0–1) was generated, and if its value was in the interval of (0–0.8), the crossover, which consisted in an arithmetic mean of the values of the parameters of the genitors, occurred, leading to a new individual. If a number between (0.8–1) was generated, the genitor with the lowest objective function was selected as the new individual of the current generation. Afterwards, another random number in the interval

(0–1) was generated. If its value was in the interval of (0–0.15), mutation occurred. The mutation consisted in increasing the parameter value resulting from the crossover step by 10% of its value. If a number between (0.15–1) was generated, no mutation occurred. For another individual, two new different genitors were randomly selected. Finally, it is important to mention that 10 independent runs were carried out using the genetic algorithm. The simplex-based model used 10^{-5} as the convergence criteria for all parameters. Table 2 presents the parameter estimation results.

From the results reported in Table 2, one can verify that the fractional order model can successfully describe the experimental dataset and it also shows to be a better model than the integer order model. This can be observed by the lower values of the objective function, the parametric variance, the narrower confidence interval and the closer to one value of the correlation coefficient presented by the fractional order model. One can conclude that this better fit occurred because the integer order model has a one parameter, while the fractional order model has two parameters. However, the value of $\sqrt{\sigma^2_{y\text{EXP}}}$ is a more neutral comparison value, as it takes into account the number of parameters of the model, this value was also smaller for the fractional order model. Finally, it is important to stress that parameter α is not a simple fitting parameter. It also shapes the mathematical function during the estimation task, according to the value of α the Mittag-Leffler function assumes a different mathematical form. As mentioned, if $\alpha = 1$, an exponential function shows up, while if $\alpha \neq 1$, another function is represented by the Mittag-Leffler expression. It is important to observe from Table 2 that the parametric correlation between k and α is approximately 0.3, which is an important result as the final value of parameter k has little influence on the final value of parameter α. According to Himmelblau [27], when the parametric correlation is close to 1 or -1, a wrong set of parameters can be estimated as a wrong value of one parameter could be compensated by a wrong value of another parameter resulting in an overall good fit. Consequently, the lower the parametric correlation, the closer to the true value the parameters tend to be. Finally, it is important to mention that the confidence interval of the fractional differential equation order, parameter α, [1.14; 1.26] does not include the value 1, therefore, it is very important to stress that the fractional derivative is statistically different from an integer order derivative.

Table 2. Parameter estimation results.

Result	Fractional Order Model	Equation	Integer Order Model	Equation
NE (Number of Experiments)	11	-	11	-
NP (Number of Parameters)	2 (k, α)	-	1 (k)	-
F_{OBJ}	312.4	(4)	1987.2	(5)
$\sqrt{\sigma^2_{\beta\text{EXP}}} = \sqrt{\frac{F_{\text{OBJ}}}{\text{NE}-\text{NP}}}$	5.89	(9)	14.09	(9)
k	3.42	-	3.11	
k (standard deviation)	0.12	(6)	0.28	(7)
k (confidence interval)	[3.14; 3.69]	(19)	[2.48; 3.75]	(20)
α	1.196	-	-	-
α (standard deviation)	0.027	(6)	-	-
α (confidence interval)	[1.14; 1.26]	(19)	-	-
$\underline{\underline{\mathbf{V}_{\text{param}}}}$ (parametric covariance matrix)	$\begin{bmatrix} 1.459 \times 10^{-2} & 9.778 \times 10^{-4} \\ 9.778 \times 10^{-4} & 7.325 \times 10^{-4} \end{bmatrix}$	(6)	$[8.170 \times 10^{-4}]$	(7)
$\underline{\underline{\mathbf{r}_{\text{param}}}}$ (parametric correlation matrix)	$\begin{bmatrix} 1 & 0.299 \\ 0.299 & 1 \end{bmatrix}$	(8)	-	-
r	0.997	(25)	0.987	(25)
r^2	0.993	-	0.972	-

Figure 1 compares the model behavior of both modeling approaches. The better fit provided by the fractional order model (Figure 1a) can be seen, as the model is closer to the experimental data when compared to the integer order model (Figure 1b). This happened as a result of the parameter estimation results listed in Table 2. Additionally, one can observe that the model confidence regions and model prediction regions are narrower in the fractional order model. As this model provided a better fit, the confidence in the model predictions tend to be closer to the true value. Finally, it is important to observe that the integer order model is an exponential function, while the fractional order model is a different function, as the value of α is different from 1. In the figures, the vertical bar errors were calculated by Equation (9) for each model, and as the value of the objective function is smaller for the fractional order model, the experimental error prediction is also smaller. The model confidence region was obtained by Equation (26) and the region of future experiment prediction was calculated by Equation (28). This last region is expected to be broader as it includes not only the model variance, but also the experimental variance.

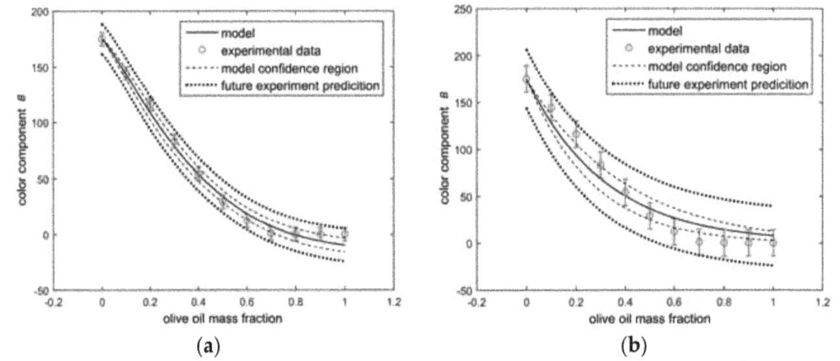

Figure 1. (**a**) Fractional order model behavior; and (**b**) integer order model behavior.

Figure 2 compares the model predictions and the experimental data of the fractional order model (Figure 2a) and the integer order model (Figure 2b). The vertical bar errors were calculated by Equation (9) and the horizontal bar errors were obtained by Equation (30). It can be seen that the model predictions of the fractional order model are closer to the experimental values:

$$B_i^{\text{EXP}} - \sqrt{\mathbf{V}_{y\text{EXP}}^{\text{MOD}}(i,i)} < B_i^{\text{EXP}} < B_i^{\text{EXP}} + \sqrt{\mathbf{V}_{y\text{EXP}}^{\text{MOD}}(i,i)}. \tag{30}$$

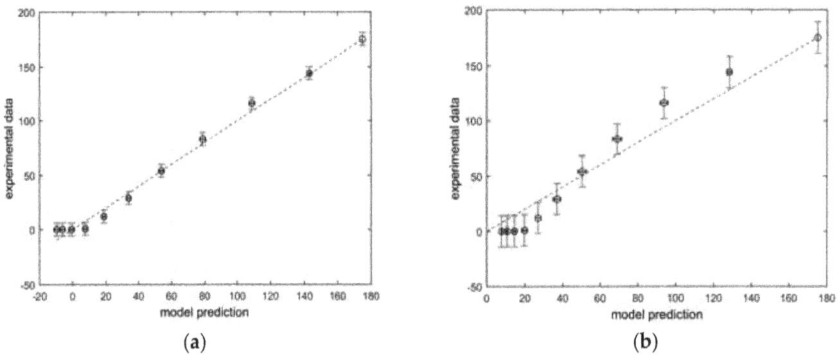

Figure 2. (**a**) Fractional order model behavior; and (**b**) integer order model behavior.

Figure 3 presents the histogram of residuals (difference between experimental data and model prediction) for the fractional order model (Figure 3a) and the integer order model (Figure 3b). It can be observed that, due to the better fit of the fractional order model, its residual histogram presents a narrower distribution and according to the mean and standard deviation, and one can conclude that the average residual is equal to zero (the standard deviation is much higher than the mean). This is an important feature, because an eventual difference may occur due to experimental error and not due to a biased model.

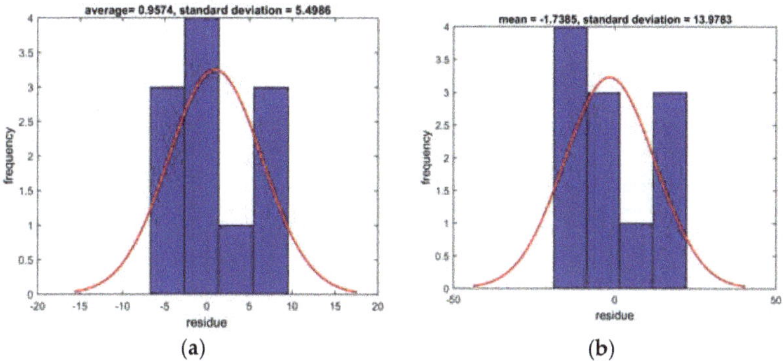

Figure 3. (**a**) Fractional order model behavior; and (**b**) integer order model behavior.

Figure 4 presents the normal probability plot of the residuals for the fractional order model (Figure 4a) and the integer order model (Figure 4b). One can observe that the residuals, besides having a mean value statistically equal to zero, can also be regarded as following a normal distribution, evidencing that the model predictions are not biased, i.e., the model does not need another term or parameter in its structure.

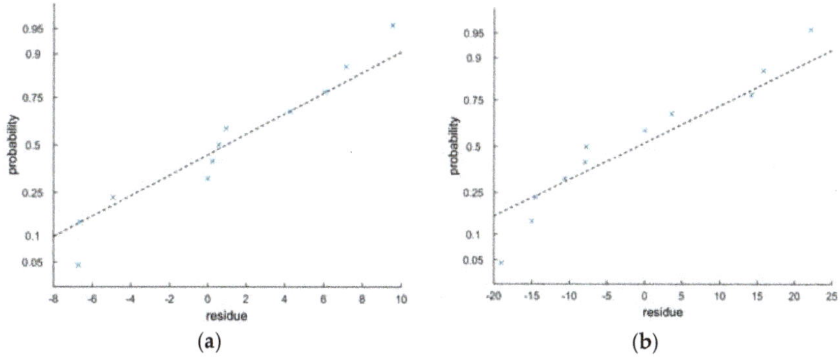

Figure 4. (**a**) Fractional order model behavior; and (**b**) integer order model behavior.

Figure 5 presents the joint confidence region calculated using the linear (Equation (23)) and nonlinear (Equation (24)) features of the objective function. The meaning of this region is that any set of parameters inside the region provides a statistically equal value of the objective function that another set of parameters would provide. Therefore, as mentioned before, it is important to stress that the values of parameter α (alpha) do not include the value 1, consequently, the value of the derivative is statistically different from an integer value. As expected, the region obtained by Equation (23) has an ellipsoidal shape and it is smaller than the region obtained by Equation (24), which considers the nonlinearities of the objective function.

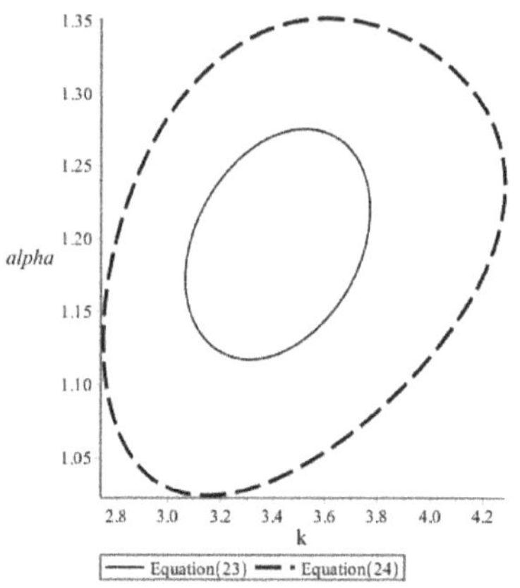

Figure 5. Joint confidence region.

Finally, Table 3 presents the model validation results. The model prediction for an olive oil mass fraction of 0.25 has a Color Component B of 93.1, which is close to the experimental value of 96. However, when considering the uncertainty of the model prediction, calculated using Equation (29), it can be regarded as statistically equal to the experimental value. The integer order model prediction is considerably different form the experimental value and the uncertainty of this prediction is even worse, which is one more indication that the fractional order model provides a better description of the experimental data and can be used for composition monitoring.

Table 3. Model Validation.

Experimental		**Fractional Order Model Prediction**	**Integer Order Model Prediction**
Olive oil mass fraction	Color Component *B*	Color Component *B*	Color Component *B*
0.25	96	93.1 ± 6.6	80.2 ± 15.2

4. Conclusions

Two different models were used to quantify different olive and soybean oil mixtures characterized by image analysis with the aid of the RGB color system. The model based on the fractional calculus-based approach could better describe the experimental dataset, presenting better results of parameter estimation quantities, such as objective function values and parameter variance. This model could successfully describe an independent validation sample, while the integer order model failed to predict the value of the validation sample. Consequently, the approach proposed here can be used as an alternative tool for possible on-line monitoring applications where a change color occurs, and it can be processed and quantified by image analysis techniques.

Acknowledgments: The authors thank the financial support and scholarships provided by CAPES and CNPQ (Brazilian Agencies).

Author Contributions: Both authors equally contributed to all aspects during the development of this work and agree with the obtained results.

Conflicts of Interest: The authors declare no conflict of interest.

References

1. Trevisan, M.G.; Poppi, R.J. Process Analytical Chemistry. *Quim. Nova* **2006**, *29*, 1065–1071. [CrossRef]
2. Russ, J.C. *The Image Processing Handbook*, 6th ed.; CRC Press: Boca Raton, FL, USA, 1972; ISBN 978-1439840450.

3. Liu, J.; Yang, W.W.; Wang, Y.S.; Rababah, T.M.; Walker, L.T. Optimizing machine vision based applications in agricultural products by artificial neural network. *Int. J. Food Eng.* **2011**, *7*, 1–23. [CrossRef]
4. Zheng, C.X.; Sun, D.W.; Zheng, L.Y. Recent applications of image texture for evaluation of food qualities—A review. *Trends Food Sci. Technol.* **2006**, *17*, 113–128. [CrossRef]
5. Litjens, G.; Kooi, T.; Bejnordi, B.E.; Setio, A.A.A.; Ciompi, F.; Ghafoorian, M.; van der Laak, J.A.W.M.; van Ginneken, B.; Sánchez, C.I. A survey on deep learning in medical image analysis. *Med. Image Anal.* **2017**, *42*, 60–88. [CrossRef] [PubMed]
6. Fernández-Caballeroa, A.; Gómeza, F.J.; López-Lópeza, J. Road-traffic monitoring by knowledge-driven static and dynamic image analysis. *Expert Syst. Appl.* **2008**, *35*, 701–719. [CrossRef]
7. Resende, M.R.; Bernucci, L.L.B.; Quintanilha, J.A. Monitoring the condition of roads pavement surfaces: Proposal of methodology using hyperspectral images. *J. Transp. Lit.* **2014**, *8*, 201–220. [CrossRef]
8. Castiñeira, D.; Rawlings, B.C.; Edgar, T.F. Multivariate Image Analysis (MIA) for Industrial Flare Combustion Control. *Ind. Eng. Chem. Res.* **2012**, *51*, 12642–12652. [CrossRef]
9. Licodiedoff, S.; Ribani, R.H.; Camlofski, A.M.O.; Lenzi, M.K. Use of image analysis for monitoring the dilution of Physalis peruviana pulp. *Braz. Arch. Biol. Technol.* **2013**, *56*, 467–474. [CrossRef]
10. Ma, F.; Hanna, M.A. Biodiesel production: A review. *Bioresour. Technol.* **1999**, *70*, 1–15. [CrossRef]
11. Moghaddam, T.B.; Baaj, H. The use of rejuvenating agents in production of recycled hot mix asphalt: A systematic review. *Constr. Build. Mater.* **2016**, *114*, 805–816. [CrossRef]
12. Labegalini, A.; Teixeira, M.L.; Ryba, A.; Villena, J. Rejuvenescimento do Ligante Asfáltico CAP 50/70 Envelhecido com Adição de Óleo de Girassol. In Proceedings of the Reunião de Pavimentação Urbana, Florianópolis, Brazil, 28–30 June 2017. (In Portuguese)
13. Tan, S.; Abraham, T.; Ference, D.; Macosko, C.W. Rigid polyurethane foams from a soybean oil-based polyol. *Polymer* **2011**, *52*, 2840–2846. [CrossRef]
14. Fernandes, J.K.; Umebara, T.; Lenzi, M.K.; Alves, E.T.S. Image analysis for composition monitoring. Commercial blends of olive and soybean oil. *Acta Sci. Technol.* **2013**, *35*, 317–324. [CrossRef]
15. Giona, M.; Roman, H.E. A theory of transport phenomena in disordered systems. *Chem. Eng. J.* **1992**, *49*, 1–10. [CrossRef]
16. Hristov, J. Multiple integral-balance method basic idea and an example with mullin's model of thermal grooving. *Therm. Sci.* **2017**, *21*, 1555–1560. [CrossRef]
17. Ionescu, C.; Lopes, A.; Copot, D.; Machado, J.A.T.; Bates, J.H.T. The role of fractional calculus in modeling biological phenomena: A review. *Commun. Nonlinear Sci.* **2017**, *51*, 141–159. [CrossRef]
18. Caratelli, D.; Mescia, L.; Bia, P.; Stukach, O.V. Fractional–Calculus–Based FDTD Algorithm for Ultrawideband Electromagnetic Characterization of Arbitrary Dispersive Dielectric Materials. *IEEE Trans. Antennas Propag.* **2016**, *64*, 3533–3544. [CrossRef]
19. Chen, W.; Sun, H.; Zhang, X.; Korošak, D. Anomalous diffusion modeling by fractal and fractional derivatives. *Comput. Math. Appl.* **2010**, *59*, 1754–1758. [CrossRef]
20. Fu, Z.-J.; Chen, W.; Yang, H.-T. Boundary particle method for Laplace transformed time fractional diffusion equations. *J. Comput. Phys.* **2013**, *235*, 52–66. [CrossRef]
21. Evangelista, L.R.; Lenzi, E.K. *Fractional Diffusion Equations and Anomalous Diffusion*, 1st ed.; Cambridge University Press: Cambrigde, UK, 2018.
22. Caputo, M. Linear models of dissipation whose Q is almost frequency independent-2. *Geophys. J. R. Astron. Soc.* **1967**, *13*, 529–538. [CrossRef]
23. Podlubny, I. Fractional-order systems and PIλDμ controllers. *IEEE Trans. Autom. Control* **1999**, *44*, 208–214. [CrossRef]
24. Isfer, L.A.D.; Lenzi, E.K.; Lenzi, M.K. Identification of biochemical reactors using fractional differential equation. *Lat. Am. App. Res.* **2010**, *40*, 193–198.
25. Lagarias, J.C.; Reeds, J.A.; Wright, M.H.; Wright, P.E. Convergence properties of the Nelder–Mead simplex method in low dimensions. *SIAM J. Optim.* **1998**, *9*, 112–147. [CrossRef]
26. Gomes, E.M.; Silva, F.R.G.B.; Araújo, R.R.L.; Lenzi, E.K.; Lenzi, M.K. Parametric Analysis of a Heavy Metal Sorption Isotherm Based on Fractional Calculus. *Math. Probl. Eng.* **2013**, 642101. [CrossRef]
27. Himmelblau, D.M. *Process Analysis by Statistical Methods*, 1st ed.; John Wiley & Sons: New York, NY, USA, 1970; ISBN 978-0471399858.

28. Bard, Y. *Nonlinear Parameter Estimation*, 1st ed.; Academic Press: New York, NY, USA, 1974; ISBN 978-0120782505.
29. Lebedev, N.N. *Special Functions & Their Applications*, 1st ed.; Dover Publications: New York, NY, USA, 1972; ISBN 978-0486606248.
30. Box, G.E.P.; Hunter, W.G. A useful method for model building. *Technometrics* **1962**, *4*, 301–318. [CrossRef]

© 2018 by the authors. Licensee MDPI, Basel, Switzerland. This article is an open access article distributed under the terms and conditions of the Creative Commons Attribution (CC BY) license (http://creativecommons.org/licenses/by/4.0/).

fractal and fractional

MDPI

Article

Stokes' First Problem for Viscoelastic Fluids with a Fractional Maxwell Model

Emilia Bazhlekova * and Ivan Bazhlekov

Institute of Mathematics and Informatics, Bulgarian Academy of Sciences, Acad. G. Bonchev Str., Bl. 8, Sofia 1113, Bulgaria; i.bazhlekov@math.bas.bg

* Correspondence: e.bazhlekova@math.bas.bg

Received: 21 September 2017; Accepted: 23 October 2017; Published: 24 October 2017

Abstract: Stokes' first problem for a class of viscoelastic fluids with the generalized fractional Maxwell constitutive model is considered. The constitutive equation is obtained from the classical Maxwell stress–strain relation by substituting the first-order derivatives of stress and strain by derivatives of non-integer orders in the interval $(0,1]$. Explicit integral representation of the solution is derived and some of its characteristics are discussed: non-negativity and monotonicity, asymptotic behavior, analyticity, finite/infinite propagation speed, and absence of wave front. To illustrate analytical findings, numerical results for different values of the parameters are presented.

Keywords: Riemann-Liouville fractional derivative; viscoelastic fluid; fractional Maxwell model; Stokes' first problem; Mittag-Leffler function; Bernstein function

1. Introduction

Viscoelastic fluid flows with the classical Maxwell constitutive model [1] have been the object of intense study for many years, for a short survey see for example [2,3]. In the case of a unidirectional flow, the Maxwell constitutive relation has the dimensionless form

$$\sigma + a\frac{\partial \sigma}{\partial t} = b\frac{\partial \varepsilon}{\partial t}, \tag{1}$$

where σ and ε are shear stress and strain, respectively, a is the relaxation time, and b the dynamic viscosity.

Fractional calculus, in view of its ability to model hereditary phenomena with long memory, has proved to be a valuable tool to handle viscoelastic aspects [4]. For instance, rheological constitutive equations with fractional derivatives play an important role in the description of properties of polymer solutions and melts [5]. A four-parameter generalization of the classical Maxwell model (1) has been proposed in [6], see also [7,8] and [9] (Chapter 7). It is obtained from (1) by substituting the first order time-derivatives of stress and strain with Riemann-Liouville fractional derivatives in time, which leads to the following generalized fractional Maxwell constitutive equation

$$\sigma + aD_t^{\alpha}\sigma = bD_t^{\beta}\varepsilon, \tag{2}$$

where $0 < \alpha, \beta \leq 1$ and $a, b > 0$. Since for $\alpha > \beta$, the corresponding relaxation function is increasing, which is not physically acceptable, the constitutive model (2) is considered only for fractional order parameters satisfying the constraints (see [6–9])

$$0 < \alpha \leq \beta \leq 1, \quad a, b > 0. \tag{3}$$

The fractional Maxwell constitutive Equation (2) with thermodynamic constraints (3) has been shown to be an excellent model for capturing the linear viscoelastic behavior of soft materials exhibiting one or more broad regions of power-law-like relaxation [8,10,11].

Stokes' first problem is one of the basic problems for simple flows. It is concerned with the shear flow of a fluid occupying a semi-infinite region bounded by a plate which undergoes a step increase of velocity from rest. One of the first studies on Stokes' first problem for viscoelastic fluids is [12]. Since then many works have been devoted to this problem for different types of fluids with linear constitutive equations. Stokes' first problem for viscoelastic fluids with general stress–strain relation is studied in [13], [14] (Section 5.4) and [4] (Chapter 4). For some recent results concerning unidirectional flows of viscoelastic fluids with fractional derivative models we refer to [15] (Chapter 7). Let us note that the fractional order governing equations are usually derived from classical models by substituting the integer order derivatives by fractional derivatives. The problem of correct fractionalization of the governing equations is discussed in [16].

The classical Maxwell constitutive Equation (1) and its fractional order generalization (2) are among the simplest viscoelastic models for fluid-like behavior and the corresponding Stokes' first problem is a good model of essential processes involved in wave propagation. For studies on Stokes' first problem for classical Maxwell fluids we refer to [2,3] and the references cited therein. Stokes' first problem for generalized fractional Maxwell fluids is studied in [17,18], where explicit solutions in the form of series expansions are obtained (in [18] only the particular case $\beta = 1$ is considered). Explicit solutions of other types of problems for unidirectional flows of fractional Maxwell fluids with constitutive Equation (2) can be found in [19–21].

Despite the abundance of works devoted to exact solutions of unidirectional flows of linear viscoelastic fluids, to the best of our knowledge, there is very little analytical work enlightening the basic properties of solution to Stokes' first problem for fractional Maxwell fluids, such as propagation speed, non-negativity, monotonicity, regularity, and asymptotic behavior. In addition, it was pointed out in [22,23] that in a number of recent works the obtained exact solutions to Stokes' first problem for non-Newtonian fluids contain mistakes. In the present work we provide a new explicit solution to Stokes' first problem for generalized Maxwell fluids with a constitutive model (2) and compare it numerically to the one given in [17]. Moreover, based on the theory of Bernstein functions, we study analytically the properties of the solution and support the analytical findings by numerical results.

This paper is organized as follows. In Section 2, the material functions of the fractional Maxwell model are studied. In Section 3, the solution to Stokes' first problem is studied analytically based on its representation in Laplace domain. Section 4 is devoted to derivation of explicit integral representation of solution and numerical computation. Some facts concerning Bernstein functions and related classes of functions are summarized in an Appendix.

2. Fractional Maxwell Model-Material Functions

Consider a unidirectional viscoelastic flow and suppose it is quiescent for all times prior to some starting time that we assume as $t = 0$. Since we work only with causal functions ($f(t) = 0$ for $t < 0$), if there is no danger of confusion for the sake of brevity, we still denote by $f(t)$ the function $H(t)f(t)$, where $H(t)$ is the Heaviside unit step function. We are concerned with the fractional Maxwell model, given by the linear constitutive Equation (2) with parameters satisfying (3). The material functions—relaxation function $G(t)$ and the creep compliance $J(t)$—in a one-dimensional linear viscoelastic model are defined by the identities [4,24]

$$\sigma(x,t) = \int_0^t G(t-\tau)\dot{\varepsilon}(x,\tau)\,\mathrm{d}\tau; \quad \varepsilon(x,t) = \int_0^t J(t-\tau)\dot{\sigma}(x,\tau)\,\mathrm{d}\tau, \ t > 0, \tag{4}$$

where the over-dot denotes the first derivative in time.

Applying (A4), we derive from the constitutive Equation (2) expressions for material functions in the Laplace domain:

$$\widehat{G}(s) = \frac{bs^{\beta-1}}{1+as^{\alpha}}, \quad \widehat{J}(s) = \frac{1}{bs^{\beta+1}} + \frac{a}{bs^{\beta-\alpha+1}}. \tag{5}$$

Taking the inverse Laplace transform of the above representations of $\widehat{G}(s)$ and $\widehat{J}(s)$, we get, by using the properties (A5) and (A8), the following representations of the material functions (see also [6–9])

$$G(t) = \frac{b}{a} t^{\alpha-\beta} E_{\alpha,\alpha-\beta+1}\left(-\frac{1}{a}t^{\alpha}\right), \quad J(t) = \frac{t^{\beta}}{b\Gamma(1+\beta)} + \frac{at^{\beta-\alpha}}{b\Gamma(1+\beta-\alpha)}, \quad t>0, \tag{6}$$

in terms of the Mittag-Leffler function (A6) and power-law type functions, respectively.

The second law of thermodynamics, stating that the total entropy can only increase over time for an isolated system, implies the following restrictions on the material functions: $G(t)$ should be non-increasing and $J(t)$ non-decreasing for $t>0$. This behavior is related to the physical phenomena of stress relaxation and strain creep [4]. It has been proven that the fractional Maxwell model (2) is consistent with the second law of thermodynamics if and only if $0<\alpha\leq\beta\leq 1$ [6–9]. For completeness, here we formulate and prove a slightly stronger statement. In fact, it appears that the monotonicity of the material functions of the fractional Maxwell model imply that the relaxation function $G(t)$ is completely monotone and the creep compliance $J(t)$ is a complete Bernstein function. The definitions and properties of completely monotone functions and related classes of functions are summarized in the Appendix.

Proposition 1. *Assume* $0<\alpha,\beta\leq 1$, $a,b>0$, $t>0$. *The following assertions are equivalent:*
(a) $0<\alpha\leq\beta\leq 1$;
(b) $G(t)$ *is monotonically non-increasing for* $t>0$;
(c) $J(t)$ *is monotonically non-decreasing for* $t>0$;
(d) $G(t)$ *is a completely monotone function;*
(e) $J(t)$ *is a complete Bernstein function.*

Proof. Using the representations (6) for the material functions, we prove that condition (a) is equivalent to any of the conditions (b)–(e). First we show that if $1\geq\alpha>\beta>0$ then (b) and (c) are not satisfied. The definition of Mittag-Leffler function (A6) implies $G(t)\sim t^{\alpha-\beta}$ for $t\to 0$. Therefore, if $\alpha>\beta$, $G(t)$ is increasing near 0, i.e., (b) is violated. Further, $\mathrm{d}J/\mathrm{d}t\sim t^{\beta-\alpha-1}/\Gamma(\beta-\alpha)$ for $t\to 0$. Since $\Gamma(\beta-\alpha)<0$ for $1\geq\alpha>\beta>0$, in this case $J(t)$ is decreasing near 0, and thus (c) is violated. Therefore, any of conditions (b) and (c) implies (a). It remains to prove that (a) implies (d) and (e), i.e., $G(t)\in\mathcal{CMF}$ and $J(t)\in\mathcal{CBF}$. Indeed, if $0<\alpha\leq\beta\leq 1$ then $t^{\alpha-\beta}\in\mathcal{CMF}$ and $E_{\alpha,\alpha-\beta+1}\left(-\frac{1}{a}t^{\alpha}\right)\in\mathcal{CMF}$ as a composition of the completely monotone Mittag-Leffler function of negative argument and the Bernstein function t^{α}. Therefore, $G(t)\in\mathcal{CMF}$ is a product of two completely monotone functions. The fact that $J(t)$ is a complete Bernstein function follows from the property of the power function $t^{\gamma}\in\mathcal{CBF}$ for $\gamma\in[0,1]$. □

Representations (6) together with the asymptotic expansion (A7) gives $G(+\infty)=0$ and $J(+\infty)=+\infty$. Therefore, the fractional Maxwell constitutive equation indeed models fluid-like behavior (for a discussion on the general conditions see e.g., [24], Section 2). More precisely, (6) and (A7) imply for $t\to+\infty$ that $G(t)\sim t^{-\alpha}$ if $\beta<1$ and $G(t)\sim t^{-\alpha-1}$ if $\beta=1$. This means that only for $\beta=1$ the relaxation function $G(t)$ is integrable at infinity and the integral over $(0,\infty)$ is finite, which is a stronger condition required for fluid behavior. In this case ($\beta=1$) we obtain from (6) and (A9) the area under the relaxation curve

$$\int_0^{\infty} G(t)\,\mathrm{d}t = b, \quad \beta=1.$$

The asymptotic expansions of the material functions for $t \to 0^+$ are $G(t) \sim t^{\alpha-\beta}$, $J(t) \sim t^{\beta-\alpha}$. Therefore, if $\alpha < \beta$ then $G(0^+) = +\infty$ and $J(0^+) = 0$, while if $\alpha = \beta$ different behavior is observed: $G(0^+) < \infty$ and $J(0^+) > 0$. More precisely, $G(0^+) = b/a = 1/J(0^+)$ for $\alpha = \beta$. This means that in the limiting case $\alpha = \beta$ the material possesses instantaneous elasticity and, therefore, finite wave speed of a disturbance is expected [4] (Chapter 4). This property will be discussed further in the next section.

3. Stokes' First Problem

Consider incompressible viscoelastic fluid which fills a half-space $x > 0$ and is quiescent for all times prior to $t = 0$. Rheological properties of the fluid are described by the fractional Maxwell constitutive Equation (2) with parameters satisfying the thermodynamic constraints (3). Taking into account that $D_t^\beta \varepsilon = {}^C D_t^\beta \varepsilon = J_t^{1-\beta} \dot{\varepsilon}$ for $\varepsilon(x,0) = 0$, see (A3), constitutive Equation (2) can be rewritten in the form:

$$(1 + aD_t^\alpha)\,\sigma(x,t) = bJ_t^{1-\beta}\dot{\varepsilon}(x,t), \;\; 0 < \alpha \le \beta \le 1, \;\; a,b > 0. \tag{7}$$

The fluid is set into motion by a sudden movement of the bounding plane $x = 0$ tangentially to itself with constant speed 1. Denote by $u(x,t)$ the induced velocity field. Assuming non-slip boundary conditions, the equation for the rate of strain $\dot{\varepsilon} = \partial u/\partial x$, Cauchy's first law $\partial u/\partial t = \partial \sigma/\partial x$ and the constitutive Equation (7), imply the following IBVP for the velocity field

$$(1 + aD_t^\alpha)\, u_t(x,t) = bJ_t^{1-\beta} u_{xx}(x,t), \;\; x,t > 0, \tag{8}$$

$$u(0,t) = H(t), \;\; u \to 0 \text{ as } x \to \infty, \;\; t > 0, \tag{9}$$

$$u(x,0) = u_t(x,0) = 0, \;\; x > 0, \tag{10}$$

where $H(t)$ is the Heaviside unit step function. Problem (8)–(10) is the Stokes first problem for a fractional Maxwell fluid, c.f. [17].

Let us note that by applying the Caputo fractional derivative operator ${}^C D_t^{1-\beta}$ to both sides of Equation (8) and taking into account the zero initial conditions (10), and operator identities (A3), Equation (8) can be rewritten as the following two-term time-fractional diffusion-wave equation

$$a\,{}^C D_t^{2+\alpha-\beta} u(x,t) + {}^C D_t^{2-\beta} u(x,t) = bu_{xx}(x,t), \;\; x,t > 0,$$

with Caputo fractional derivatives of orders $2 + \alpha - \beta, 2 - \beta \in [1,2]$. In a number of works various problems for the two-term time-fractional diffusion-wave equation are studied, e.g., [25], [26] (Chapter 6) and [27–32]. In this work, we study the Stokes' first problem following the technique developed in [33] for the multi-term time-fractional diffusion-wave equation.

Problem (8)–(10) is conveniently treated using Laplace transform with respect to the temporal variable. Denote by $\widehat{u}(x,s)$ the Laplace transform of $u(x,t)$ with respect to t. By applying Laplace transform to Equation (8) and taking into account the boundary conditions (9), initial conditions (10) and identities (A4), the following ODE for $\widehat{u}(x,s)$ is obtained

$$g(s)\widehat{u}(x,s) = \widehat{u}_{xx}(x,s), \; \widehat{u}(0,s) = 1/s, \; \widehat{u}(x,s) \to 0 \text{ as } x \to \infty, \tag{11}$$

where

$$g(s) = \frac{s}{\widehat{G}(s)} = \frac{as^{2+\alpha-\beta}}{b} + \frac{s^{2-\beta}}{b}. \tag{12}$$

Solving problem (11), with s considered as a parameter, it follows, c.f. [17] (Equation (22))

$$\widehat{u}(x,s) = \frac{1}{s}\exp\left(-x\sqrt{g(s)}\right), \tag{13}$$

where $g(s)$ is defined in (12).

Before taking inverse Laplace transform to obtain explicit representation of the solution, we first deduce information about its behavior directly from its Laplace transform (13).

First we prove that $\sqrt{g(s)}$ is a complete Bernstein function. Here this is deduced from the complete monotonicity of the relaxation function $G(t)$ proved in Proposition 1. Therefore, the proposed method of proof can be used also for more general models, provided the corresponding relaxation function is completely monotone.

Proposition 2. *Assume* $0 < \alpha \leq \beta \leq 1, a, b > 0$. *Then* $\sqrt{g(s)} \in \mathcal{CBF}$.

Proof. For the proof we use properties (D), (E) and (G) in the Appendix. Since $G(t) \in \mathcal{CMF}$ then $\widehat{G}(s) \in \mathcal{SF}$ by definition. Therefore, property (D) implies $1/\widehat{G}(s) \in \mathcal{CBF}$. Since also $s \in \mathcal{CBF}$ (by (G)), $g(s) = s/\widehat{G}(s)$ is a product of two complete Bernstein functions, and (E) implies $\sqrt{g(s)} \in \mathcal{CBF}$. □

Theorem 1. *The solution of Stokes' first problem* $u(x,t)$ *is monotonically non-increasing function in* x *and monotonically non-decreasing function in* t*, such that* $0 \leq u(x,t) \leq 1$ *and* $u(x, 0^+) = 0$, $u(x, +\infty) = 1$. *More precisely, for any* $x > 0$,

$$u(x,t) \sim 1 - \frac{x t^{\beta/2-1}}{\sqrt{b}\Gamma(\beta/2)}, \quad t \to +\infty. \tag{14}$$

Proof. From Proposition 2 and properties (A), (B) and (C) in the Appendix we deduce (in the same way as in [33] (Theorem 2.2)):

$$\exp\left(-x\sqrt{g(s)}\right) \in \mathcal{CMF}, \ \frac{1}{s}\exp\left(-x\sqrt{g(s)}\right) \in \mathcal{CMF}, \ \frac{\sqrt{g(s)}}{s}\exp\left(-x\sqrt{g(s)}\right) \in \mathcal{CMF}. \tag{15}$$

Therefore, (13) and Bernstein's theorem imply

$$u(x,t) \geq 0, \ \frac{\partial}{\partial t}u(x,t) \geq 0, \ -\frac{\partial}{\partial x}u(x,t) \geq 0, \quad x, t > 0, \tag{16}$$

and, in this way, the non-negativity and monotonicity properties of the solution are proven. Taking into account that $g(s) \to 0$ as $s \to 0$ and $g(s) \to \infty$ as $s \to \infty$, it follows from (13) and the final and initial value theorem of Laplace transform

$$\lim_{t\to 0^+} u(x,t) = \lim_{s\to\infty} s\widehat{u}(x,s) = 0, \quad \lim_{t\to+\infty} u(x,t) = \lim_{s\to 0} s\widehat{u}(x,s) = 1, \tag{17}$$

and, therefore, $0 \leq u(x,t) \leq 1$. Since $\exp\left(-x\sqrt{g(s)}\right) \sim 1 - x\sqrt{g(s)}$ and $g(s) \sim s^{-\beta}/b$ for $s \to 0$, then (13) implies

$$\widehat{u}(x,s) \sim \frac{1}{s} - x\frac{s^{-\beta/2}}{\sqrt{b}}, \quad s \to 0,$$

which, by applying (A5) and Tauberian theorem, gives the asymptotic expansion (14) of $u(x,t)$. □

Next, propagation speed of a disturbance is discussed. From general theory, see e.g., [4] (Chapter 4) and [14] (Chapter 5), the velocity of propagation of the head of the disturbance is $c = \kappa^{-1}$, where

$$\kappa = \lim_{s\to\infty} \frac{\sqrt{g(s)}}{s} = \lim_{s\to\infty} \frac{1}{\sqrt{s\widehat{G}(s)}} = \lim_{t\to 0^+} \frac{1}{\sqrt{G(t)}}.$$

From here, we easily obtain the propagation speed for the considered fractional Maxwell model

$$c = \begin{cases} \infty, & 0 < \alpha < \beta \leq 1; \\ \sqrt{b/a}, & 0 < \alpha = \beta \leq 1. \end{cases}$$

Therefore, if $0 < \alpha < \beta \leq 1$, the disturbance propagates with infinite speed. On the other hand, if the orders of the two derivatives in the model are equal $0 < \alpha = \beta \leq 1$ (and thus the model possesses instantaneous elasticity as discussed earlier), the propagation speed c is finite. This means that $u(x,t) \equiv 0$ for $x > ct$. Let us consider the case of finite propagation speed in more detail. It is known that in the limiting case of classical Maxwell model ($\alpha = \beta = 1$) the velocity field $u(x,t)$ has a jump discontinuity at the planar surface $x = ct$, see e.g., [2]. What happens when the orders α and β are equal, but non-integer? According to [14] the amplitude of such a jump is $\exp(-\omega x)$, where $\omega = \lim_{s\to\infty}\left(\sqrt{g(s)} - \kappa s\right)$. In our specific case (12) with $\alpha = \beta$ gives $g(s) = \frac{as^2}{b} + \frac{s^{2-\beta}}{b} = \kappa^2 s^2\left(1 + \frac{1}{as^\beta}\right)$ and using the expansion $\sqrt{1+x} \sim 1 + x/2$ for $x \to 0$, we find

$$\omega = \lim_{s\to\infty} \kappa s\left(\sqrt{1 + \frac{1}{as^\beta}} - 1\right) = \lim_{s\to\infty} \frac{\kappa}{2a} s^{1-\beta}. \tag{18}$$

Therefore, $\omega < \infty$ only in the case $\beta = 1$, which means that only in the case of classical Maxwell fluid the wave $u(x,t)$ exhibits a wave front, i.e., a discontinuity at $x = ct$. For equal non-integer values of the parameters α and β, $0 < \alpha = \beta < 1$, (18) implies $\omega = \infty$, and thus there is no wave front at $x = ct$. Therefore, in this case, an interesting phenomenon is observed: coexistence of finite propagation speed and absence of wave front. This is a memory effect, which is not present for linear integer order models.

At the end of this section we consider in more detail the case $0 < \alpha < \beta \leq 1$, for which we obtained $c = \infty$. We will prove next that in this case the solution $u(x,t)$ is an analytic function in t.

Theorem 2. *Let $0 < \alpha < \beta \leq 1$ and $\theta_0 = \frac{(\beta-\alpha)\pi}{2(2+\alpha-\beta)} - \varepsilon$, where $\varepsilon > 0$ is arbitrarily small. Then, for any $x > 0$, the solution $u(x,t)$ of Stokes' first problem (8)-(9)-(10) admits analytic extension to the sector $|\arg t| < \theta_0$, which is bounded on each sub-sector $|\arg t| \leq \theta < \theta_0$.*

Proof. It is sufficient to prove that $\hat{u}(x,s)$ admits analytic extension to the sector $|\arg s| < \pi/2 + \theta_0$ and $s\hat{u}(x,s)$ is bounded on each sub-sector

$$|\arg s| \leq \pi/2 + \theta, \quad \theta < \theta_0, \tag{19}$$

see e.g., [14] (Theorem 0.1). Indeed, since $\sqrt{g(s)} \in \mathcal{CBF}$, by property (H) in the Appendix, it can be analytically extended to $\mathbb{C}\backslash(-\infty, 0]$ and, hence, the same will hold for $\hat{u}(x,s)$. Moreover, from (12) we deduce for any s satisfying (19)

$$\left|\arg\sqrt{g(s)}\right| \leq \frac{2+\alpha-\beta}{2}|\arg s| \leq \pi/2 - \varepsilon_0,$$

where $\varepsilon_0 = \frac{2+\alpha-\beta}{2}\varepsilon \in (0, \pi/2)$. Inserting this inequality in (13), we obtain for any $x > 0$ and any s satisfying (19) the estimate $|s\hat{u}(x,s)| \leq \exp\left(-x|g(s)|^{1/2}\sin\varepsilon_0\right) \leq 1$. □

The analyticity of the solution established in the above theorem implies that for any $x > 0$ the set of zeros of $u(x,t)$ on $t > 0$ can be only discrete. This, together with the monotonicity of $u(x,.)$ proved in Theorem 1, implies that $u(x,t) > 0$ for all $x, t > 0$. This observation confirms again that in the case $0 < \alpha < \beta \leq 1$ a disturbance spreads infinitely fast.

4. Explicit Representation of Solution and Numerical Results

To find explicit representation of the solution we apply the Bromwich integral inversion formula to (13) and obtain the following integral in the complex plane

$$u(x,t) = \frac{1}{2\pi \mathrm{i}} \int_{D \cup D_0} \frac{1}{s} \exp\left(st - x\sqrt{g(s)}\right) \mathrm{d}s, \tag{20}$$

where the Bromwich path has been transformed to the contour $D \cup D_0$, where

$$D = \{s = \mathrm{i}r,\ r \in (-\infty, -\varepsilon) \cup (\varepsilon, \infty)\},\ D_0 = \{s = \varepsilon e^{\mathrm{i}\theta}, \theta \in [-\pi/2, \pi/2]\}.$$

In the same way as in [33] (Theorem 2.5) we deduce from (20) the following real integral representation of the solution.

Theorem 3. *The solution of Stokes' first problem (8)–(10) admits the integral representation:*

$$u(x,t) = \frac{1}{2} + \frac{1}{\pi} \int_0^\infty \exp(-xK^+(r)) \sin(rt - xK^-(r)) \frac{\mathrm{d}r}{r}, \quad x,t > 0, \tag{21}$$

where

$$K^{\pm}(r) = \frac{1}{\sqrt{2}} \left(\left(A^2(r) + B^2(r) \right)^{1/2} \pm A(r) \right)^{1/2}$$

with

$$A(r) = -\frac{a}{b} r^{2+\alpha-\beta} \cos \frac{(\beta-\alpha)\pi}{2} - \frac{1}{b} r^{2-\beta} \cos \frac{\beta\pi}{2}, \quad B(r) = \frac{a}{b} r^{2+\alpha-\beta} \sin \frac{(\beta-\alpha)\pi}{2} + \frac{1}{b} r^{2-\beta} \sin \frac{\beta\pi}{2}.$$

Let us check that the integral in (21) is convergent. Indeed, since $K^-(r) \sim r^{1-\beta/2}$ as $r \to 0$ then $\sin(rt - xK^-(r))/r \sim r^{-\beta/2}$ as $r \to 0$. Therefore, the function under the integral sign in (21) has an integrable singularity at $r \to 0$. At $r \to \infty$ integrability is ensured by the term $\exp(-xK^+(r))$, since $K^+(r) > 0$ and $K^+(r) \sim r^{1-(\beta-\alpha)/2} \to +\infty$ as $r \to +\infty$.

Next, the explicit integral representation (21) is used for numerical computation and visualization of the solution to the Stokes' first problem for different values of the parameters. For the numerical computation of the improper integral in (21) the MATLAB subroutine "integral" is used. The aim of our numerical computations is two-fold: to support the presented analytical findings and to compare our results to those in [17] (see their Figures 1 and 2), where plots of the solution are given, obtained from a different representation: a series expansion in terms of Fox H-functions. In order to compare our numerical solutions to the ones given in [17], we perform computations for the same parameters. In contrast to [17], where only plots of solution as a function of x are given, we plot it also as a function of t.

In Figure 1 the solution $u(x,t)$ is plotted for $\beta = 0.8$ and four different values of α: $0.2, 0.4, 0.6$, and 0.8; $a = 10^\alpha$, $b = 10^{\beta-1}$. The case of instantaneous elasticity $\alpha = \beta = 0.8$ is added to those plotted in Figure 1 of [17], in order to show the different behavior. In this case, the disturbance at $x = 0$ propagates with finite speed $c = \sqrt{b/a} = 10^{-1/2}$ and $u(x,t) \equiv 0$ for $x > ct$. There is no wave front at $x = ct$. In the other three cases with $\alpha < \beta$ the propagation speed is infinite and $u(x,t)$ should be positive for all $x, t > 0$. However, far from the wall, it becomes negligible. As expected, the solution $u(x,t)$ exhibits monotonic behavior: it is non-increasing with respect to x, see Figure 1a, and non-decreasing with respect to t, see Figure 1b. The plots presented in Figure 1a for $\alpha = 0.2, 0.4, 0.6$ are identical with those given in Figure 1 of [17].

In Figure 2, the solution is plotted for $\alpha = 0.5$ and three different values of β: $0.5, 0.7, 0.9$, $a = 10^\alpha$, $b = 10^{\beta-1}$. The case of instantaneous elasticity now is $\alpha = \beta = 0.5$ with finite propagation speed c having the same value as above. Again, as expected, there is no wave front at $x = ct$. Monotonicity of

$u(x,t)$ is clearly seen. The plots presented in Figure 2a for fixed time $t = 4$ are identical with those given in [17], Figure 2.

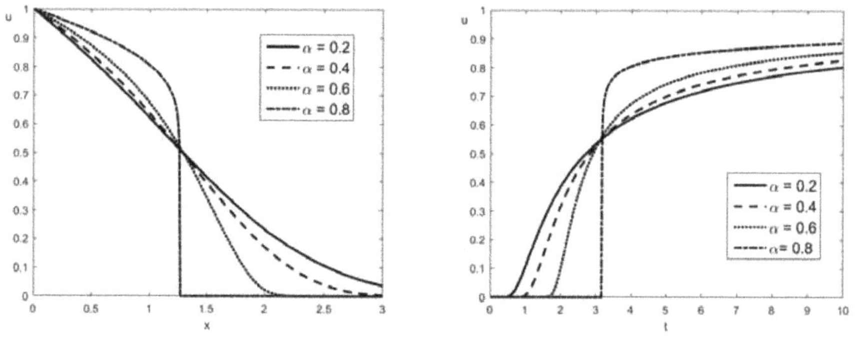

(a) (b)

Figure 1. Velocity field $u(x,t)$ for $\beta = 0.8$ and different values of α: 0.2, 0.4, 0.6, and 0.8; $a = 10^{\alpha}$, $b = 10^{\beta-1}$; (**a**) $u(x,t)$ as a function of x for $t = 4$; (**b**) $u(x,t)$ as a function of t for $x = 1$.

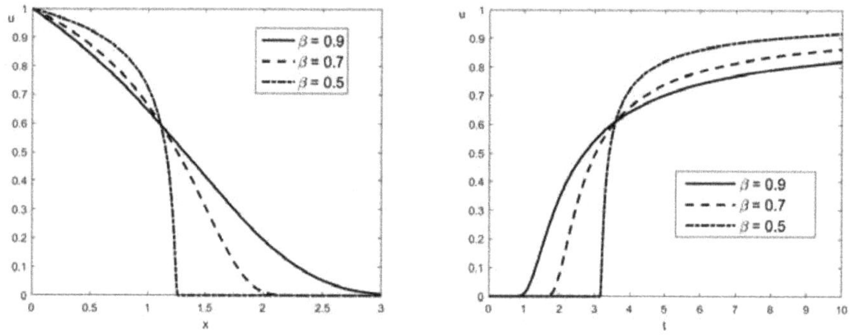

(a) (b)

Figure 2. Velocity field $u(x,t)$ for $\alpha = 0.5$ and three different values of β: 0.5, 0.7, 0.9; $a = 10^{\alpha}$, $b = 10^{\beta-1}$; (**a**) $u(x,t)$ as a function of x for $t = 4$; (**b**) $u(x,t)$ as a function of t for $x = 1$.

We refer also to [33] (Figures 1–3) for more illustrations of the three different types of behavior, related to Maxwell fluids, namely: finite wave speed and presence of wave front ($\alpha = \beta = 1$), finite wave speed and absence of wave front ($0 < \alpha = \beta < 1$), and infinite propagation speed ($0 < \alpha < \beta \leq 1$). Let us note that the paper [18] also presents plots of the solution of Stokes' first problem for fractional Maxwell fluids with $\beta = 1$. However, the plots in Figures 5a, 7a and 9a of [18] do not seem to be monotonically decreasing with respect to the spatial variable, which is a contradiction with the expected monotonic behavior of the solution.

5. Conclusions

The solution of Stokes' first problem for a viscoelastic fluid with the generalized Maxwell model with fractional derivatives of stress and strain is studied based on its Laplace transform with respect to the temporal variable. The thermodynamic constraints on the fractional parameters imply physically reasonable monotonic behavior of solution with finite propagation speed (without wave front) when the two fractional orders coincide and infinite propagation speed otherwise. An explicit integral representation of the solution is derived and used for its numerical computation. The obtained numerical results are in agreement with those reported in [17].

The explicit integral representation of the solution can be further used to derive explicit or asymptotic expressions for other characteristics of the flow, such as shear stress at the plate, thickness of the boundary layer, etc.

The presented technique can be applied in the study of different generalizations of the considered model, e.g., the viscoelastic model based on Bessel functions in [34] and the fractional order weighted distributed parameter Maxwell model in [35].

Acknowledgments: This work is supported by Bulgarian National Science Fund (Grant DFNI-I02/9); and performed in the frames of the bilateral research project between Bulgarian and Serbian academies of sciences "Analytical and numerical methods for differential and integral equations and mathematical models of arbitrary (fractional or high integer) order".

Author Contributions: Both authors contributed to the conception and development of this work.

Conflicts of Interest: The authors declare no conflict of interest.

Appendix A

The Riemann-Liouville and the Caputo fractional derivatives, D_t^α and ${}^C D_t^\alpha$, are defined by the identities

$$D_t^\alpha = \frac{\mathrm{d}^m}{\mathrm{d}t^m} J_t^{m-\alpha}, \quad {}^C D_t^\alpha = J_t^{m-\alpha} \frac{\mathrm{d}^m}{\mathrm{d}t^m}, \quad 0 \le m-1 < \alpha \le m,\ m \in \mathbb{N}, \tag{A1}$$

where J_t^β denotes the Riemann-Liouville fractional integral:

$$J_t^\beta f(t) = \frac{1}{\Gamma(\beta)} \int_0^t (t-\tau)^{\beta-1} f(\tau)\,\mathrm{d}\tau = \left\{\frac{t^{\beta-1}}{\Gamma(\beta)}\right\} * f, \ \beta > 0; \quad J_t^0 = I; \tag{A2}$$

with $*$ being the Laplace convolution

$$(f * g)(t) = \int_0^t f(t-\tau) g(\tau)\, d\tau.$$

Some basic properties of the fractional order operators are listed next:

$$\begin{aligned} &J_t^\alpha J_t^\beta = J_t^{\alpha+\beta}, \ {}^C D_t^\alpha J_t^\alpha = D_t^\alpha J_t^\alpha = I \ \ \forall \alpha, \beta > 0; \\ &{}^C D_t^\alpha f = D_t^\alpha f, \ J_t^\alpha {}^C D_t^\alpha f = J_t^\alpha D_t^\alpha f = f \text{ if } f^{(k)}(0) = 0,\ k = 0, 1, ..., m-1,\ m-1 < \alpha \le m. \end{aligned} \tag{A3}$$

Denote the Laplace transform of a function by

$$\mathcal{L}\{f(t)\}(s) = \widehat{f}(s) = \int_0^\infty e^{-st} f(t)\,\mathrm{d}t.$$

The Laplace transform of fractional order operators obeys the following identities, where $\alpha > 0$:

$$\begin{aligned} &\mathcal{L}\{J_t^\alpha f\}(s) = s^{-\alpha} \mathcal{L}\{f\}(s); \\ &\mathcal{L}\{D_t^\alpha f\}(s) = \mathcal{L}\{{}^C D_t^\alpha f\}(s) = s^\alpha \mathcal{L}\{f\}(s), \ \text{ if } f^{(k)}(0) = 0,\ k = 0, 1, ..., m-1,\ m-1 < \alpha \le m, \end{aligned} \tag{A4}$$

which can be derived from the definitions (A1), (A2), the Laplace transform pair

$$\mathcal{L}\left\{\frac{t^{\beta-1}}{\Gamma(\beta)}\right\} = s^{-\beta}, \ \beta > 0, \tag{A5}$$

the convolution property $\mathcal{L}\{(f * g)(t)\}(s) = \widehat{f}(s)\widehat{g}(s)$, and the identity for the integer order derivative $\mathcal{L}\{f^{(m)}\}(s) = s^m \mathcal{L}\{f\}(s)$ if $f^{(k)}(0) = 0$ for $k = 0, 1, ..., m-1$.

The two-parameter Mittag-Leffler function is defined by the series

$$E_{\alpha,\beta}(z) = \sum_{k=0}^{\infty} \frac{z^k}{\Gamma(\alpha k+\beta)}, \quad \alpha,\beta,z \in \mathbb{C},\ \Re\alpha > 0. \tag{A6}$$

It admits the asymptotic expansion

$$E_{\alpha,\beta}(-t) = \frac{t^{-1}}{\Gamma(\beta-\alpha)} - \frac{t^{-2}}{\Gamma(\beta-2\alpha)} + O(t^{-3}),\ \alpha \in (0,2), \beta \in \mathbb{R},\ t \to +\infty, \tag{A7}$$

and satisfies the Laplace transform identity

$$\mathcal{L}\left\{t^{\beta-1}E_{\alpha,\beta}(-\lambda t^{\alpha})\right\}(s) = \frac{s^{\alpha-\beta}}{s^{\alpha}+\lambda}. \tag{A8}$$

The following relation is often useful:

$$\frac{\mathrm{d}}{\mathrm{d}t}E_{\alpha,1}\left(-\frac{1}{a}t^{\alpha}\right) = -\frac{1}{a}t^{\alpha-1}E_{\alpha,\alpha}\left(-\frac{1}{a}t^{\alpha}\right). \tag{A9}$$

For details on Mittag-Leffler functions we refer to [36,37].

A function $\varphi : (0,\infty) \to \mathbb{R}$ is said to be completely monotone function ($\varphi \in \mathcal{CMF}$) if it is of class C^{∞} and

$$(-1)^n \varphi^{(n)}(x) \geq 0,\ \lambda > 0, n = 0,1,2,... \tag{A10}$$

The characterization of the class $\mathcal{CMF}$ is given by the Bernstein's theorem: a function is completely monotone if and only if it can be represented as the Laplace transform of a non-negative measure (non-negative function or generalized function).

The class of Stieltjes functions ($\mathcal{SF}$) consists of all functions defined on $(0,\infty)$ which can be written as Laplace transform of a completely monotone function. Therefore, $\mathcal{SF} \subset \mathcal{CMF}$.

A non-negative function φ on $(0,\infty)$ is said to be a Bernstein function ($\varphi \in \mathcal{BF}$) if $\varphi'(x) \in \mathcal{CMF}$; $\varphi(x)$ is said to be a complete Bernstein functions ($\mathcal{CBF}$) if and only if $\varphi(x)/x \in \mathcal{SF}$. We have the inclusion $\mathcal{CBF} \subset \mathcal{BF}$.

A selection of properties of the classes of functions defined above is given next.

(A) The class $\mathcal{CMF}$ is closed under point-wise multiplication.
(B) If $\varphi \in \mathcal{BF}$ then $\varphi(x)/x \in \mathcal{CMF}$.
(C) If $\varphi \in \mathcal{CMF}$ and $\psi \in \mathcal{BF}$ then the composite function $\varphi(\psi) \in \mathcal{CMF}$.
(D) $\varphi \in \mathcal{CBF}$ if and only if $1/\varphi \in \mathcal{SF}$.
(E) If $\varphi, \psi \in \mathcal{CBF}$ then $\sqrt{\varphi.\psi} \in \mathcal{CBF}$.
(F) The Mittag-Leffler function $E_{\alpha,\beta}(-x) \in \mathcal{CMF}$ for $0 < \alpha \leq 1, \alpha \leq \beta$.
(G) If $\alpha \in [0,1]$ then $x^{\alpha} \in \mathcal{CBF}$ and $x^{\alpha-1} \in \mathcal{SF}$;
(H) If $\varphi \in \mathcal{SF}$ or $\mathcal{CBF}$ then it can be analytically extended to $\mathbb{C}\backslash(-\infty,0]$ and $|\arg\varphi(z)| \leq |\arg z|$ for $z \in \mathbb{C}\backslash(-\infty,0]$.

For proofs and more details on these special classes of functions we refer to [38].

References

1. Maxwell, J. On the dynamical theory of gasses. *Phil. Trans. R. Soc. Lond.* **1867**, *157*, 49–88.
2. Jordan, P.M.; Puri, A. Revisiting Stokes' first problem for Maxwell fluids. *Q. J. Mech. Appl. Math.* **2005**, *58*, 213–227.
3. Jordan, P.M.; Puri, A.; Boros, G. On a new exact solution to Stokes' first problem for Maxwell fluids. *Int. J. Non Linear Mech.* **2004**, *39*, 1371–1377.
4. Mainardi, F. *Fractional Calculus and Waves in Linear Viscoelasticity*; Imperial College Press: London, UK, 2010.

5. Bagley, R.L.; Torvik, P.J. On the fractional calculus model of viscoelastic behavior. *J. Rheol.* **1986**, *30*, 137–148.
6. Friedrich, C. Relaxation and retardation functions of the Maxwell model with fractional derivatives. *Rheol. Acta* **1991**, *30*, 151–158.
7. Schiessel, H.; Metzler, R.; Blumen, A.; Nonnenmacher, T.F. Generalized viscoelastic models: Their fractional equations with applications. *J. Phys. A* **1995**, *28*, 6567–6584.
8. Makris, N.; Dargush, G.F.; Constantinou, M.C. Dynamic analysis of generalized viscoelastic fluids. *J. Eng. Mech.* **1993**, *119*, 1663–1679.
9. Hilfer, R. *Applications of Fractional Calculus in Physics*; World Scientific: Singapore, 2000.
10. Hernandez-Jimenez, A.; Hernandez-Santiago, J.; Macias-Garcia, A.; Sanchez-Gonzalez, J. Relaxation modulus in PMMA and PTFE fitting by fractional Maxwell model. *Polym. Test.* **2002**, *21*, 325–331.
11. Jaishankar, A.; McKinley, G.H. A fractional K-BKZ constitutive formulation for describing the nonlinear rheology of multiscale complex fluids. *J. Rheol.* **2014**, *58*, 1751–1788.
12. Tanner, R.I. Note on the Rayleigh problem for a visco-elastic fluid. *Z. Angew. Math. Phys.* **1962**, *13*, 573–580.
13. Preziosi, L.; Joseph, D.D. Stokes' first problem for viscoelastic fluids. *J. Non Newtonian Fluid Mech.* **1987**, *25*, 239–259.
14. Prüss, J. *Evolutionary Integral Equations and Applications*; Birkhäuser: Basel, Switzerland, 1993.
15. Zheng, L.; Zhang, X. *Modeling and Analysis of Modern Fluid Problems*; Academic Press: Cambridge, MA, USA, 2017.
16. Hristov, J. Emerging issues in the Stokes first problem for a Casson fluid: From integer to fractional models by the integral–balance approach. *J. Comput. Complex. Appl.* **2017**, *3*, 72–86.
17. Tan, W.; Xu, M. Plane surface suddenly set in motion in a viscoelastic fluid with fractional Maxwell model. *Acta Mech. Sin.* **2002**, *18*, 342–349.
18. Jamil, M.; Rauf, A.; Zafar, A.A.; Khan, N.A. New exact analytical solutions for Stokes' first problem of Maxwell fluid with fractional derivative approach. *Comput. Math. Appl.* **2011**, *62*, 1013–1023.
19. Yang, D.; Zhu, K.Q. Start-up flow of a viscoelastic fluid in a pipe with a fractional Maxwell's model. *Comput. Math. Appl.* **2010**, *60*, 2231–2238.
20. Yin, Y.; Zhu, K.Q. Oscillating flow of a viscoelastic fluid in a pipe with the fractional Maxwell model. *Appl. Math. Comput.* **2006**, *173*, 231–242.
21. Tan, W.; Pan, W.; Xu, M. A note on unsteady flows of a viscoelastic fluid with the fractional Maxwell model between two parallel plates. *Int. J. Non Linear Mech.* **2003**, *38*, 645–650.
22. Christov, I.C. On a difficulty in the formulation of initial and boundary conditions for eigenfunction expansion solutions for the start-up of fluid flow. *Mech. Res. Commun.* **2013**, *51*, 86–92.
23. Christov, I.C. Comments on: Energetic balance for the Rayleigh–Stokes problem of an Oldroyd-B fluid [Nonlinear Anal. RWA 12 (2011) 1]. *Nonlinear Anal.* **2011**, *12*, 3687–3690.
24. Mainardi, F.; Spada, G. Creep, relaxation and viscosity properties for basic fractional models in rheology. *Eur. Phys. J. Spec. Top.* **2011**, *193*, 133–160.
25. Atanacković, T.M.; Pilipović, S.; Zorica, D. Diffusion wave equation with two fractional derivatives of different order. *J. Phys. A* **2007**, *40*, 5319–5333.
26. Atanacković, T.M.; Pilipović, S.; Stanković, B.; Zorica, D. *Fractional Calculus with Applications in Mechanics: Vibrations and Diffusion Processes*; John Wiley & Sons: London, UK, 2014.
27. Mamchuev, M.O. Solutions of the main boundary value problems for the time-fractional telegraph equation by the Green function method. *Fract. Calc. Appl. Anal.* **2017**, *20*, 190–211.
28. Qi, H.; Guo, X. Transient fractional heat conduction with generalized Cattaneo model. *Int. J. Heat Mass Transf.* **2014**, *76*, 535–539.
29. Chen, J.; Liu, F.; Anh, V.; Shen, S.; Liu, Q.; Liao, C. The analytical solution and numerical solution of the fractional diffusion-wave equation with damping. *Appl. Math. Comput.* **2012**, *219*, 1737–1748.
30. Qi, H.T.; Xu, H.Y.; Guo, X.W. The Cattaneo-type time fractional heat conduction equation for laser heating. *Comput. Math. Appl.* **2013**, *66*, 824–831.
31. Bazhlekova, E. On a nonlocal boundary value problem for the two-term time-fractional diffusion-wave equation. *AIP Conf. Proc.* **2013**, *1561*, 172–183.
32. Bazhlekova, E. Series solution of a nonlocal problem for a time-fractional diffusion-wave equation with damping. *C. R. Acad. Bulg. Sci.* **2013**, *66*, 1091–1096.

33. Bazhlekova, E.; Bazhlekov, I. Subordination approach to multi-term time-fractional diffusion-wave equations. *arXiv* **2017**, arXiv:1707.09828.
34. Colombaro, I.; Giusti, A.; Mainardi, F. A class of linear viscoelastic models based on Bessel functions. *Meccanica* **2017**, *52*, 825–832.
35. Cao, L.; Li, Y.; Tian, G.; Liu, B.; Chen, Y.Q. Time domain analysis of the fractional order weighted distributed parameter Maxwell model. *Comput. Math. Appl.* **2013**, *66*, 813–823.
36. Gorenflo, R.; Kilbas, A.; Mainardi, F.; Rogosin, S. *Mittag-Leffler Functions, Related Topics and Applications*; Springer: Berlin/Heidelberg, Germany, 2014.
37. Paneva-Konovska, J. *From Bessel to Multi-Index Mittag–Leffler Functions: Enumerable Families, Series in Them and Convergence*; World Scientific: Singapore, 2016.
38. Schilling, R.L.; Song, R.; Vondraček, Z. *Bernstein Functions: Theory and Applications*; De Gruyter: Berlin, Germany, 2010.

© 2017 by the authors. Licensee MDPI, Basel, Switzerland. This article is an open access article distributed under the terms and conditions of the Creative Commons Attribution (CC BY) license (http://creativecommons.org/licenses/by/4.0/).

MDPI
St. Alban-Anlage 66
4052 Basel
Switzerland
Tel. +41 61 683 77 34
Fax +41 61 302 89 18
www.mdpi.com

Fractal Fract Editorial Office
E-mail: fractalfract@mdpi.com
www.mdpi.com/journal/fractalfract

www.ingramcontent.com/pod-product-compliance
Lightning Source LLC
Chambersburg PA
CBHW051910210326
41597CB00033B/6096
9783038429838